To Amanda,
WARMLY!

Jerry ~ Kathy

Beacon Blankets

Make Warm Friends

Jerry and Kathy Brownstein

4880 Lower Valley Road, Atglen, PA 19310 USA

Dedication

This book is dedicated to the many generations and the thousands of fine men and women, devoted loyal Beacon employees, whose high standards and superior work ethic expressed the confidence that Beacon blankets do indeed make warm friends; and to the Charles D. Owens and their families, who believed in these people and the quality products they produced together.

Meeting at the Grove Park Inn, Asheville, North Carolina, Summer 2000. Pictured left to right: Tedd Smith, Vice President of Sales, Owen Manufacturing Co.; George Lemieux, retired Plant General Manager, Beacon Manufacturing Co.; Dan Owenby, previous Vice President, Product Development, Beacon Manufacturing Co. and presently at Owen Manufacturing Co.; John W. Hill, previous Senior Vice President of Corporate Planning, Beacon Manufacturing Co. ; Jerry Brownstein, author; and Charles D. Owen III, President, Owen Manufacturing Co.

Designed by John P. Cheek
Cover design by Bruce M. Waters
Type set in Humanist 521 BT/Souvenir Lt BT

ISBN: 0-7643-1359-2

Printed in China

Published by Schiffer Publishing Ltd.
4880 Lower Valley Road
Atglen, PA 19310
Phone: (610) 593-1777; Fax: (610) 593-2002
E-mail: Schifferbk@aol.com
Please visit our web site catalog at
www.schifferbooks.com
We are always looking for people to write books on new and related subjects. If you have an idea for a book, please contact us at the above address.

This book may be purchased from the publisher. Include $3.95 for shipping. Please try your bookstore first. You may write for a free catalog.

In Europe, Schiffer books are distributed by
Bushwood Books
6 Marksbury Ave. Kew Gardens
Surrey TW9 4JF England
Phone: 44 (0)20 8392-8585; Fax: 44 (0)20 8392-9876
E-mail: Bushwd@aol.com
Free postage in the UK. Europe: air mail at cost.
Please try your bookstore first.

Acknowledgments

A grateful and heartfelt "Thank you" to all our contributors, especially John W. Hill, George Lemieux, Dan Owenby, Charles D. Owen Jr. IV, Chairman and CEO of Owen Manufacturing Company, and his son, Charles D. Owen III, President of Owen Manufacturing Company.

John W. Hill spent thirty four years of his professional life at the Beacon Manufacturing Co. His career began in accounting/data processing and he rose through the ranks becoming senior vice president of Manufacturing Services, working in New York Monday through Friday and then working in Swannanoa on Saturday and Sunday. John was promoted again in 1994 to senior vice president of Corporate Planning over seeing the entire Beacon operation until 1997. He provided us his generous and outstanding assistance and we have developed a lasting friendship.

George Lemieux worked forty seven years for Beacon, starting in 1931 at age fourteen as an errand boy in the New Bedford plant while his father, a thirty year employee, was a supervisor in the weave room. George was a bright and eager young man and arranged to secretly learn how to "cut cards" (the punch cards that directed the Jacquard machines through their operations) personally paying for instruction late at night. One day after the company moved to Swannanoa, a job opened for a card cutter and George's father announced that his son was trained and available for the position. George quickly became an expert in complex blanket making machinery. He served his country during World War II from 1942-1945. Upon his return he was named supervisor in the weave room. He worked his way up the ladder and was sent all over the world to find and purchase machinery for the Beacon operation. When Beacon acquired Esmond Mills in Canada in the late 1940s, George went there for three months. In the 1950s he was off to several Latin American countries to set up factories that would make blankets part of the year and sugar sacks the rest of the year, utilizing Beacon's retired machinery. He is the father of the "needle woven" blanket, and it is estimated that Mr. Lemieux was involved in acquiring 90% of the manufacturing equipment at Beacon. George has proudly stated, "Everyone (at Beacon) worked together as one big family." His last four years before retirement in 1978 were as plant General Manager. He is still available for advise. We sincerely appreciate the depth of knowledge that he shared with us.

Dan Owenby, formerly vice president of Product Development and currently president of Beacon's Quarter Century Club, was at Beacon for thirty nine years. Dan grew up in the Beacon village and graduated from Charles Owen High School. He was the bat boy for the Beacon baseball team and officially started working for Beacon in 1961 loading box cars at night for $1.12 per hour. He worked his way up through several positions until an opening came in the design room under Howard Rogers and George Lemieux. Dan is highly regarded as a talented design engineer with a thorough understanding of what a machine can and cannot do. Dan commented "Beacon provided an environment in which a person *could* grow." His contributions and help for this book, and his personal encouragement, were immeasurable. Dan currently directs product development at Owen Manufacturing Company.

Special thanks to historian C. Dexter Schierenbeck, grandson of Charles Owen Dexter; William E. Berner; Tedd Smith; Ken McElrath; Pat Harris; Drew Ansley; David Hollywell; Tammy Thomas; Jack McMahan, Charles D. Owen Middle School; Ben Talbert, Charles D. Owen High School; Judy Hudson and Norwood Barns, Bearwallow Mountain Traders, Hendersonville, North Carolina; Rod Dyer; Kerrie Quinn; John and Cindy Speare, RetroViva Seattle, Washington and Portland, Oregon; Brenda Cain, Antiques and Textiles, Santa Monica, California; Laura Fisher/Antique Quilts & Americana, New York, New York; R. Greg Otto, Portland, Oregon; Gary Holt and Steve Christianson, Gary Holt Collections, Missoula, Montana; David W. Schutzler, Enumclaw, Washington; Paul & Stephanie Hauger, Yellowstone Vintage Clothing Co., Santa Barbara & Hollywood, California; Sam and Denise Kennedy, Cisco's, Coeur d'Alene, Idaho; Marvin Leib; Jane Ross, Sierra Hills Antiques, Grass Valley, California; Carolyn Bartlett, Pine Lodge Rustic Cabins, Black Mountain, North Carolina; Connie & Dave Filkins, Puyallup, Washington; Cotton Incorporated, Amy Gerdnick, New York; Deb Barney, Lands End, Coming Home. Libraries: Asheville Buncombe Library System, Asheville North Carolina; King County Library System, Issaquah, Washington; Pack Memorial Library, North Carolina Collection, Asheville, North Carolina; and Seattle Public Library, Seattle main branch, Seattle, Washington. University of North Carolina at Asheville, Public Information Office, Jill Yarnall. Professional photography: Sudden Image, Randy Ledford, Fletcher, North Carolina. A special thanks to Debbie Dean and Kate Bradley who willingly and graciously helped with editing and Peter and Nancy Schiffer our publisher and editor respectively for seeing the necessity for this book. Thank you everyone, we couldn't have done it with out your help.

A Norman Rockwell oil painting commissioned by Charles D. Owen II. *The Ombre plaid blanket*, c. 1927. Courtesy of Charles D. Owen Jr. IV.

Contents

Preface

Evolution of The Beacon Manufacturing Company or The Tale of Two Companies

The Alamo Company, a joint venture of Phillips 66 Petroleum Company and National Distillers & Chemical Corporation, purchased 50 percent of Beacon stock from the Owen family in 1964. The Federal Trade Commission stepped in and broke up the "budding conglomerate" and National Distillers & Chemical Corporation emerged with the Beacon stock. In 1969 National purchased the remaining stock from the Owen family and took complete control of Beacon. This ended four generations of the Owen family at Beacon. Charles D. Owen Jr. IV started his own blanket manufacturing company in the early 1970s, The Charles D. Owen Manufacturing Co., not far from the Beacon plant in Swannanoa.

Beacon was next acquired by Cannon Mills in 1981, specifically for its blanket operation. They operated the company until Cannon Mills and its subsidiaries were purchased by David H. Murdock (of Dole Pineapple Company).

The subsequent sale of Cannon Mills to Fieldcrest Mills in early 1986 did not include Beacon, as Fieldcrest had its own blanket division. Beacon remained owned by Mr. Murdock's Pacific Holding Company until 1994 when Pillowtex Corporation of Dallas, Texas purchased the company.

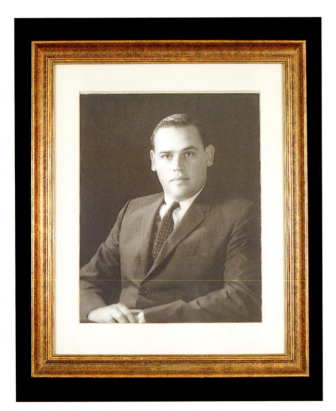

Charles D. Owen Jr. IV, 1935, founder, Chairman, and C.E.O. of Owen Manufacturing Company.

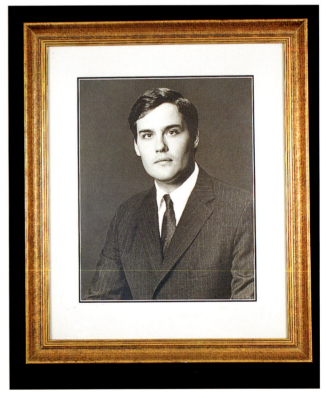

Charles D. Owen V, (III) 1960, president of Owen Manufacturing Company. Charles D. Owen V renumbered himself III. Courtesy of Owen Manufacturing Co.

Chapter I

The Beacon Is Lit: Company History

All cotton blankets are not Beacon brand and Beacon brand blankets are not all cotton (as some believe), but most are. The Beacon Manufacturing Company has been the world's largest producer of all types of blankets for almost a century and their name has become synonymous with patterned cotton blankets regardless of which company may have manufactured them.

In The Beginning

The Charles D. Owen families are members of the Society of Mayflower Descendants. They trace their origins through marriage to John Alden, Richard Warren, Isaac Alerton, and John Howland, all original Mayflower passengers. The first Owens to arrive in America were Samuel and his wife, Priscilla, from Wales, in 1651. They located in Providence, Rhode Island, where the family grew and prospered.

The first Charles Dexter Owen (1841-1915) was descended from a long line of successful business people. His father was a well known and respected gold jewelry manufacturer in Providence, Rhode Island. In 1866 Charles D. Owen was employed by the Charles H. Clark Company, wool dealers. In 1876 he was working in worsted yarns, and by 1878 he had joined with his brother in a textile venture. 1880 saw Charles D. Owen as an agent for the Atlantic Mills. In 1881 he became managing director of St. Croix Cotton Mills in Mill Town, New Brunswick, Canada, where his cousin, Charles Owen Dexter, came to work in 1885. The cousins worked together again at Colored Cotton Mills Company Ltd of Hamilton and Merritton, Ontario, Canada. In 1892, Owen returned to Atlantic Mills as director and treasurer.

Pictured on the left and seated, Fallee (Fally) Palmer Owen (Mrs. George Owen), 1816-1911, great-great-grandmother of Charles D. Owen Jr. IV. Standing, Charles D. Owen I, 1841-1915. Seated with baby, Charles D. Owen II, 1875-1937. The baby is Charles D. Owen Sr. III, 1904-1990. Courtesy of Charles D. Owen Jr. IV.

Charles D. Owen I, 1841-1915, co-founder of Beacon Manufacturing Company.

Charles D. Owen I was a serious sailing competitor and a member of the New York Yacht Club. His schooner, The Sachem, which means "chief" in some North American Indian languages, was a stunning craft.

Mr. Owen and his crew won this challenge October 2, 1886 by eight minutes, due to favorable winds and a superior craft. Courtesy of Charles D. Owen Jr. IV.

Charles Owen Dexter, 1862-1943, Beacon Manufacturing Company co-founder, designer, and inventor.

His son, Charles D. Owen II (1875-1937), graduated from Brown University in 1897 and went to Europe to study the most modern processes in wool blanket manufacturing.

Atlantic Mills, Providence, Rhode Island, July 6, 1897. Courtesy of Charles D. Owen Jr. IV.

Charles D. Owen II,1857-1937, Beacon Manufacturing Company co-founder.

In 1904, father and son joined with their cousin Charles Owen Dexter, a textile engineer with expertise in cotton fabrics, to acquire the defunct New Bedford, Massachusetts, company, The Beacon Manufacturing Company, that had manufactured yarns from waste. They purchased the vacant 35,000 square foot building along with its charter and name. Seventy-eight state-of-the-art looms, twenty of them Jacquards, were shipped to the new owners from Europe. The Jacquard loom, invented by Joseph Marie Jacquard (1752-1834) of France, was capable of weaving intricate designs using up to four colors.

Charles D. Owen II, third from left, with his executive staff, "c. 1918." Courtesy of Owen Manufacturing, Co.

The Beacon manufacturing Company, New Bedford, Massachusetts. Courtesy of Charles D. Owen Jr. IV.

The new Beacon Manufacturing Company started production with twenty employees in March of 1905. The earliest product line was cotton flannel fabric which was finished into bathrobes, house robes, and regular bed blankets. Charles D. Owen I personally sold Beacon's products during the first year of operation.

Product acceptance was so strong that in just seven years, by 1912, the work force was 16 times larger and the company employed 800 persons. The product lines were expanded to include infant blankets and bathrobe piece goods which were sold to other manufactures.

In 1915 Charles D. Owen I died and his son, Charles II, took control of the company. By the end of 1919 the company had again expanded many times and had become the largest blanket manufacturing company under one roof in the United States.

Beacon Manufacturing Company outing complete with an employees' band at Perry's Grove, 1920. Courtesy of Owen Manufacturing, Co.

A Vertical Company

From its start, Beacon was a vertical operation. They bought raw materials and did all the converting, spinning, dying, and finishing within their own facility. The company had its own laboratory for testing dyes for color fastness and, as a result, Beacon cotton blankets had the enviable reputation of being washable and resistant to the colors' fading or running.

Reprocessing cotton waste, a by-product of manufacturing, required elaborate machinery. Beacon took the initiative and was able to recover and reuse much of its own waste materials. Beacon also had its own maintenance departments and could repair or make parts for the hundreds of different machines that operated throughout the plant.

From its beginning Beacon adhered to a policy of directly selling from its mill to retailers. Soon after the establishment of a sales office in New York another office was opened in Chicago, with additional offices being added in most major cities. This method of selling caused Beacon to carry enormous inventories in its storehouse as orders were filled from stock. It took a large capital reserve to keep Beacon going and growing.

Products being produced in the 1920s included crib blankets, single and double blankets, Indian design blankets, robe flannel cloth, institutional wool blankets and yarns for sale to other manufacturers.

Many private label programs were produced for major retail companies including J.C. Penny, Montgomery Ward and Sears, Roebuck & Co. Additional retailers included W.T. Grant Company and dry good stores in the United States and around the world.

A New Era Emerges

Worldwide respect had developed for Beacon blankets and the product line was considered totally different from anything in its field. What was unique about Beacon fabric was the successful development of their process to treat cotton as if it were wool. An ordinary cotton manufacturer's objective was to get the individual fibers to lay nearly parallel to each other. Beacon's objective was to have as few parallel fibers as possible. They wanted soft, fluffy, lofty yarn resulting in air pockets which created a warmer blanket. Beacon had identified which cotton raw material was key to the fluffiest and fuzziest end product. This raw material was known as "short staple cotton" and the earliest supplies came from China.

The entire Beacon line of blankets continued to grow steadily and rapidly in consumer acceptance and esteem. A big boost to cotton blankets came just prior to 1920, when the U.S. Bureau of Standards compared qualities and weights of cotton and wool blankets. The tests showed that Beacon cotton blankets far exceeded the government standard for heat retaining qualities compared to army and/or navy wool blankets. The Bureau announced that, "No one need doubt that cotton blankets are indeed fully as warm as wool." This endorsement, along with Beacon's stylish designs, color combinations and attention to detail were additional reasons for the vogue that this bed covering company established. At this point there was no doubt that Beacon blankets had taken over the market and the drab, single color blankets were becoming a thing of the past.

Midway Prizes

The *Sunday Standard Newspaper* of New Bedford, Massachusetts, on September 30, 1923, reported, "Beacon blankets have instant and perennial appeal to the everyday American citizen which is illustrated by their popularity for prizes on the Midway in various fairs that are held throughout the country."

MIDWAY BLANKETS — Singles

MIDWAY 37

MIDWAY 38

STYLE MIDWAY.   Single.   Size 60 x 80.   Ends hemmed.

COLORS: MIDWAY 37		COLORS: MIDWAY 38	
Red and Blue	Blue and Red	Tan and Brown	Green and Dark Green
Rust and Green	Tan and Brown	Rose and Maroon	Wine and Green
Green and Red	Wine and Grey	Blue and Red	Black and Red
		Lavender and Dark Lavender	

(Page 13)

Midway blankets from the 1939 catalog.

Fakirs, the term used for the people who made a business of conducting Midway attractions, declared, "There is nothing that compares with the Beacon blanket in its suitability for prizes and that these blankets have the same business pulling power in all parts of the country."

In 1923, after fewer than twenty years in business, the company capacity had multiplied fifteen times with the plant covering more than thirteen acres. The storehouse covered approximately 200,000 square feet. There were 1600 employees and 1200 looms, mostly Jacquard, and, as one observer noted, "No two (looms) were doing the same thing at the same time!"

Swannanoa

In 1923, Charles Owen II was traveling by train through the spectacular hill country near Asheville, North Carolina, on the look-out for an additional factory location. He spotted a level tract of land in Swannanoa, just east of Asheville, in Buncombe County. (Historians believe that the definition of "Swannanoa" is "beautiful valley".) He thought it would be a perfect site for a factory.

The purchase was arranged. It included 160 acres of farm land, 12 of which were level. The factory was built on the level ground, where it is to this day. Over time, Beacon purchased over 1300 acres from the top of the mountain to the Swannanoa river valley.

Ground breaking for the new location was in 1924, and the plant was ready for operations in 1925. Twenty-five percent of the New Bedford equipment had been shipped from New Bedford by train and the new plant started by producing limited specialty products. Construction went on for another ten years.

Many of the people who helped build the plant stayed on to work there; for example, Andy Whitson, son of the farmer who sold the original site, became chief of piping and plumbing maintenance. The main street to the plant, Whitson Avenue, is named after his family.

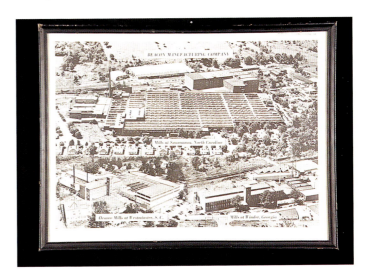

The Southern Beacon Mills, late 1930.

11

Beacon Employees' Sick Relief Association parade, "c. 1930."
Courtesy of Beacon Manufacturing Co.

Owen's Folly

Taking an enormous risk in the beginning of the Great Depression, Charles D. Owen, Sr. (III) (1904-1990) responded to economic changes in Massachusetts by suggesting to his father that the entire company be moved south to the new plant location in Swannanoa. The area was close to raw materials, electricity was cheaper and there was an abundance of bright and eager workers. Consequently, plans were made to move the "mother plant" to the "beautiful valley". Executives of other textile companies called Beacon's move "Owen's Folly", but in time they were to change their minds and followed, resulting in "the greatest textile migration in history" and the establishment of the southern textile industry.

Charles D. Owen Sr. III, 1904-1990.

Federal Trade Commission and Navajo Indians vs. Beacon Manufacturing Company

Right in the middle of all the moving plans, the Federal Trade Commission launched a complaint against Beacon. The complaint was filed on Nov. 3, 1930 by the F.T.C. and the Navajo Indian Nation and the final order with the decision came on June 28, 1932. The Commission found that the company's display and names of blankets were injurious to the genuine Navajo Indian weavers and conveyed a false impression to the buying public.

An adult Indian and child weaving a blanket together. Courtesy of Beacon Manufacturing Co..

12

The Beacon Manufacturing Company of New Bedford, Massachusetts and Swannanoa, North Carolina, was ordered by the Federal Trade Commission to stop using Indian pictures and Indian names in its advertisements and window display materials unless accompanied by an explanation that the blankets were not Indian made.

Under the cease and desist order, Beacon could continue to call its products "Indian" blankets, provided it used qualifying phrases as "Beacon Manufacturing Company Indian blankets", "Beacon Indian design blankets", or "Indian design blankets". The order specifically prohibited the use of pictures of Indians weaving blankets, looms on which Indian blankets are made, or Indian camping and marriage scenes that included Beacon blankets as part of the picture. Color pictures of a child guided by an elderly Indian in weaving a blanket and an adult Indian squatting at a loom were particularly objected to. The commission stated, "The representation by means of pictures that the blankets are made by Indians is as complete as if made by adequate words, and is more impressive and lasting."

Beacon was given sixty days to file a report with the commission to detail the manner and form which they would conform to. Beacon complied with the order.

The Big Move

In 1933 the largest blanket company in the world began to move in its entirety from New Bedford. It was a massive undertaking, requiring thousands of railroad freight cars. Not only were all the machinery and inventory moved, but many Beacon employees and their families decided to relocate as well. Plant executives, managers and skilled workers who were well trained in designing and creating these beautiful blankets followed the shift south. Many of these very skilled and talented employees would have been difficult to replace had they remained in the North.

New England had been a melting pot of immigrants from Europe, and many of these people were Beacon employees. So, with the migration came the introduction of many different nationalities, including French, Polish, German, Greek, Portuguese, Irish, Italian, Spanish and one family from the Azores. The hill country now had new names on the mailboxes and in the phone book, names like Lemieux, Swistack, Astley, Lazotte, Oliver, Fontaine and Vailancourt.

Beacon had built a village near the new plant that included a company store for groceries and clothing, a bank, a gas station and various other stores. There was company housing available with very reasonable rents. As time went by, the village businesses were purchased and operated privately. Eventually, the homes were offered to the employees to purchase at very reasonable prices.

1934 Executive photograph taken at the Beacon plant in Swannanoa. First row, left to right, Mr. Shore, Mr. Vincent, Mr. Blecher, Mr. Daly. Second row, Mr. Deal, Mr. Morgan, Mr. Magnant, Mr. Gibbs, Mr. Mallory. Third row, Mr. Smith, Mr. Cleveland, Mr. Young, Mr. Powell, Mr. Edwards, Mr. Vailancourt. Courtesy of Beacon Manufacturing Co.

The entire movement and subsequent set up were completed by about 1936, and the New Bedford facility was history. The Swannanoa plant was now the largest blanket mill in the world with over one million square feet.

Two views inside the massive Swannanoa plant.

13

By the late 1930s there were fifteen hundred styles in their line and Beacon reported sales of twenty-one million blankets while maintaining an inventory of two million blankets in stock. Despite the enormity of the new mill, Beacon continued to acquire additional support facilities in South Carolina, Georgia and Canada in the late thirties and forties.

Potluck party at the plant, 1940. Courtesy of Beacon Manufacturing.

Beacon Swannanoa, the largest blanket mill in the world, producing over twenty-one million blankets a year.

In 1937, Beacon experienced its second loss in executive management with the death of Charles D. Owen II. The cloak of responsibility was then passed on to Charles D. Owen III, who became Chairman of the Board and Treasurer.

1942-1945, The War Years

In 1942, at the start of U.S. involvement in World War II, 930 people from Beacon's work force volunteered for military service. This was remarkable, considering that the total work force was 2,200 at the time. It was the greatest contribution of workers made by any single industrial company in America. New employees had to be hired and trained to replace those who had volunteered for service.

Beacon, along with several other companies, was asked by the government to bid on the production of 100% wool blankets for all branches of the armed services. Machines were converted from cotton and cotton/wool production, and scarce parts were made in Beacon's own shops. Six months later the plant was ready to go. The company produced seven million top quality blankets for the armed forces during the war years, using about twenty-five million pounds of premium wool.

The first Christmas dinner, Spinning, Filling, and Warps area, December, 1939. Courtesy of Beacon Manufacturing Co.

Winter 1946, Executive photograph with Southern Railroad representatives. Left to right on locomotive, top to bottom, Bob Dardner from the railroad, Frank Vincent, Bill Johnson, Jade Shore, Charles Edwards, Alden Gifford, Hal Cory, all with Beacon. Mr. Lawrence and Mr. Oliver from the railroad. Courtesy of Beacon Manufacturing Co.

War Memorials

The World War I memorial plaque was moved to Swannanoa from New Bedford and is mounted on the reverse side of the large World War II monument, located near the front of the plant's main office on Whitson Avenue.

World War I Memorial Plaque dedication.

The names of Beacon employees who served in WWI. Names with *stars* beside them gave their lives for our country.

World War II Monument Plaque Dedication.

Open Door Policy

The roll-top desk that has been used by each generation of the Charles D. Owen family, from the beginning of the Beacon Manufacturing Company until its' current use at the Charles D. Owen Manufacturing Company, was well known to thousands of Beacon employees. Many policy decisions were made at this desk, yet the most far-reaching of these was the "open door" policy. It demonstrated management's sincere interest in the employees who could bring their work-related grievances or personal problems directly to Mr. Owen and receive help. This policy created harmony and loyalty from the employees. Part of Beacon's success is attributed to its "open door" policy.

The original Owen roll top desk. Courtesy of Charles D. Owen Jr. IV.

Owen Family Legacy

Throughout the Owen family's history in Swannanoa and Buncombe County, a genuine interest for their community and their employees existed. The Owens have shared in many causes and demonstrated countless acts of kindness and appreciation, supporting an impressive array of community projects.

The names of Beacon employees who served in WWII. Names with *stars* beside them gave their lives for their country.

The Charles and Catherine Owen Heart Center at Mission St. Joseph Hospital, in Asheville, was named in their honor in recognition of their exceptional service.

On the campus of the University of North Carolina, Asheville, stands Owen Hall and Conference Center which, when dedicated in 1980, was the largest classroom building in the university's history.

The middle school in Swannanoa was originally the high school and is named for Charles D. Owen II. In 1991 a new high school was built and named for Charles D. Owen, Sr. III. Charles D. Owen park, on a beautiful site near the Owen Manufacturing Company plant, was donated by Charles D. Owen Jr. IV.

The dedication plaque at the middle school (originally the high school); portrait of Charles Owen II hangs above the plaque.

The sign at the entrance of the new high school.

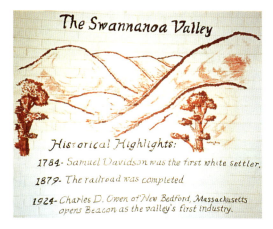

Historical mural at the middle school (formerly the high school).

Dedication plaque at the new high school.

The Owen Pool, located near Owen Park.

High school fund raising blanket/throw, great for football games.

Quarter Century Club

Each year Beacon sponsors an employee dinner party for the Quarter Century Club. Members must have at least twenty five years of employment at Beacon to be invited. In 1999 over three hundred members attended.

Beacon Baseball Teams

The Beacon team played baseball in the Textile Manufacturing League from the 1930s to 1960s. Beacon's team was considered very competitive and it was rumored that some people were hired just to play baseball. Beacon also sponsored a Little League team around the same time.

White horse mascot and the new Owen High School, partially seen on the right.

The Beacon baseball team, 1940. Courtesy of Beacon Manufacturing.

Home of The Beacon Blankets

Young Ben Talbert arrived in Swannanoa with his family from Akin, South Carolina where a big billboard had always proclaimed a welcome that read, "Akin, South Carolina, Home of the Akin Hornets", in honor of the high school football team. Upon Ben's arrival in Swannanoa he noticed a similar sign, "Swannanoa, North Carolina, Home of (the) Beacon Blankets", and he thought to himself, "What a strange name for a football team!"

Today Mr. Talbert is vice principal at the Charles D. Owen High School, Home of the Warhorses & Warlassies, and has a chuckle every time he thinks about that billboard.

Dan Owenby, whose family members had worked their entire careers at Beacon, started his Beacon experience at a very young age as the baseball team's bat boy. Dan owned a dog by the name of "Watch" who would find lost balls that he would later "redeem" for fifty cents each. Dan was vice president of Product Development and had been at Beacon 39 years before joining Owen Manufacturing Company in late 2000.

Chapter 2
Blanket Designs and Original Artwork

The Artists and Their Art

E.I. Couse

Charles D. Owen II was an avid patron of Native American art. He acquired many paintings by Eanger Irving Couse (1866-1936), especially those in which Indian blankets were represented. The art had a distinct impact on Beacon blanket designs and was used in promotional materials such as cutouts for window displays and in advertising materials. Charles D. Owen II shared many of Mr. Couse's views about the plight of Native Americans. Several of Mr. Couse's paintings remain in the Beacon archives.

Indian father demonstrating blanket making to his son; E.I. Couse artist. Courtesy of Beacon Manufacturing Co.

Mr. Couse was born in Saginaw, Michigan. He visited local settlements of Chippewa and Ogibwa tribes to observe and sketch these Indians in their daily routines and habitat. He studied at the Art Institute of Chicago and at National Academy of Design in New York, but it was during his four years in Paris studying under Adolphe Bouguereau at the Academie Julian that he developed the French Academic painting style that was reflected in his work through-out his career.

Indian wedding painting; E.I. Couse, artist. Courtesy of Beacon Manufacturing Co.

An Indian father and son admiring a blanket.

18

E. I. Couse married fellow student Virginia Walker in Paris. Her home was in the state of Washington and in 1891 they came to live near the Klikitat, Yakima and Umatilla tribes, where he produced some of his early Indian paintings. Mr. Couse visited Taos in 1902 and realized that this area was everything he had dreamed of, combining the art world with his deep interest in Indians and their artifacts. He first spent the summers in Taos and moved there permanently in 1927. His Indian paintings for the Atchison, Topeka and Santa Fe Railway Company gained him serious recognition as an artist. He was founder and first president of the Taos Society of Artists.

Indian displaying a blanket. Courtesy of Beacon Manufacturing Co.

Indian woman with her papoose wrapped in a blanket. The words "Beacon Blankets" can be seen faintly on her skirt. Courtesy of Beacon Manufacturing Co.

Young Indian on a hill top wearing blue beaded moccasins and lying on a blanket, watching the scene below. Courtesy of Beacon Manufacturing Co.

Indian admiring a blanket. Courtesy of Beacon Manufacturing Co..

An Indian child wrapped in a blanket. Courtesy of Beacon Manufacturing Co.

Advertising cutout of an Indian and boy, admiring a piece of Beacon robe cloth (fabrics could be changed to suit the display).Courtesy of Beacon Manufacturing Co.

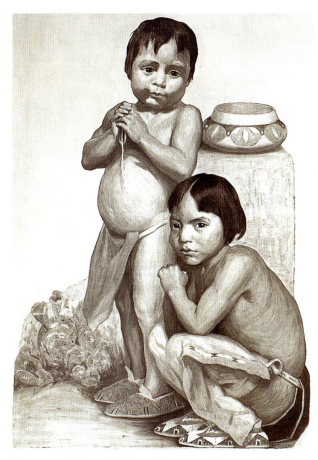

Two young Indian children. Courtesy of Beacon Manufacturing Co.

R. Farrington Elwell

R. Farrington Elwell (1874-1962), artist, cowboy and rodeo contestant, produced many paintings and illustrations for Charles D. Owen II. Mr. Elwell was a friend of Buffalo Bill Cody and spent summers on his Wyoming ranch, the famous TE.

In 1896 Cody offered Elwell the position of ranch manager, which he held for the next twenty-five years. The surroundings inspired his art work. Sioux Chief Iron Tail deemed him an honorary member of the Sioux tribe and his friends included Frederic Remington and Annie Oakley.

A painting by R. Farrington Elwell showing an Indian woman weaving a blanket while being observed by her husband. Courtesy of Beacon Manufacturing Co.

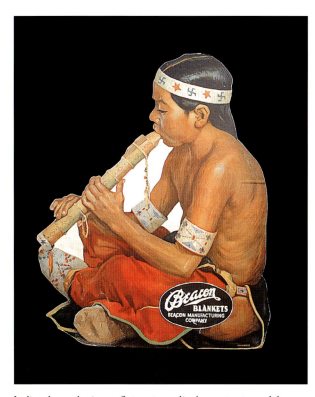

Indian boy playing a flute, store display cutout used for advertising. Courtesy Beacon Manufacturing Co.

Norman Rockwell

Additionally, Norman Rockwell (1894-1978), famous for his *Saturday Evening Post* magazine covers, was commissioned by Mr. Owen to do several paintings and drawings for use in Beacon catalogs, brochures and store window displays. Rockwell's realistic style worked well with the image and atmosphere Beacon wanted to create with their blankets. Mr. Rockwell lived in and around the northeast for most of his life and spent his final years in Stockbridge, Massachusetts.

Elder seaman with Beacon *Ombre* blanket on his lap, showing a model of a sailing ship to his grandson; artist, Norman Rockwell. Oil on canvas. Courtesy of Charles D. Owen Jr. IV.

Box Covers

An original painting possibly made for an Indian blanket box cover. Artist unknown. Courtesy of Beacon Manufacturing Co.

Comfortables (Jacquard) box cover. "c. 1920s."

Indian blankets box cover. "c. 1920s."

Designers

Charles Owen Dexter

Charles O. Dexter (1862-1943) was a Brown University graduate, class of 1885, and co-founder of the Beacon Manufacturing Company, along with his cousin, Charles Dexter Owen I, and Owen's son, Charles Dexter Owen II. The Owens and Dexter are the fathers of the Beacon cotton blanket. As an inventor, Dexter developed the *Ombre* design process and was responsible for perfecting colorfast vegetable dyes. A genius improviser, his contributions were numerous. He was blind in one eye from a household accident and had a glass eye replacement. George Lemieux, a friend and co-worker remarked, "Mr. Dexter could see more with one eye than most men could with two".

Mr. Dexter drove daily from his 123 acre estate, Shawme Farm,in Sandwich, Cape Cod to the New Bedford mill. He was a master horticulturist, specializing in hybrid rhododendrons. Not wanting to leave his estate when the company moved to Swannanoa, he consulted down south. After he passed away in 1943, his Beacon stock was purchased from his wife by the Owens. The E.I. Lilly Foundation purchased the estate several years later and Mr. Dexter's work became part of his legacy.

William H. Berner

William H. Berner (1892-1967), Head Designer, designed Beacon fabrics for fifty-three years from 1914 until he died on the job in 1967. He was responsible for designing the entire product line of blankets, robe cloth, baby blankets and blanket programs for major retailers. His inspiration for Indian designs came from the Indian art collection that Charles D. Owen II had acquired.

Mr. Berner's office was located in the New York showroom until 1941 when he, his family and the art department, moved to Swannanoa.

Mr. Berner's art and grasp of consumer's taste were key factors in Beacon's success.

William H. Berner, Head Designer at his art table. He spent fifty three years at Beacon designing America's top selling blankets. Courtesy of Charles D. Owen III.

William E. Berner

W. H. Berner's son, William E. Berner, started working at the mill part time in 1947. He became the first management trainee in 1955 and climbed the corporate ladder, working in a variety of "front office" financial positions. Finally reaching vice-president of Forecasting, he retired in 1996 after forty one years with the company. Between father and son, they spent a total of 94 years with Beacon.

This kind of loyalty was well known within the company as many generations of families worked proudly for Beacon.

Howard Rogers

Howard Rogers (1910-1987), Chief Designer, Beacon Plant Swannanoa, was a self-taught artist whose speciality was designing and adapting blanket designs for weaving on Jacquard looms. He was considered a genius for his ability to solve complex design manufacturing problems. The department he supervised covered ten acres and had nine hundred looms working three shifts a day.

Howard Rogers, Chief Designer, worked at Swannanoa for forty one years, helping to produce millions of colorful Beacon blankets. Courtesy of Dan Owenby.

Mr. Rogers started his career at Beacon in 1932, when he was twenty-two years old. He learned everything he could about blanket manufacturing, as he put it, "from the dye house to the sample room." He spent forty-one years at Beacon, producing millions of blankets during that time. Mr. Rogers and the Jacquard looms were retired at the same time, in 1973.

Design Concepts

Blanket and robe cloth designs were first created as watercolor paintings. If the design was approved, punch cards would be made to direct the elaborate and complex weaving equipment in creating the pattern.

Note the grid pattern lined in on the paintings for this purpose. Several paintings have notations indicating the year and the company the patterns were being designed for. The first green design concept is from 1927.

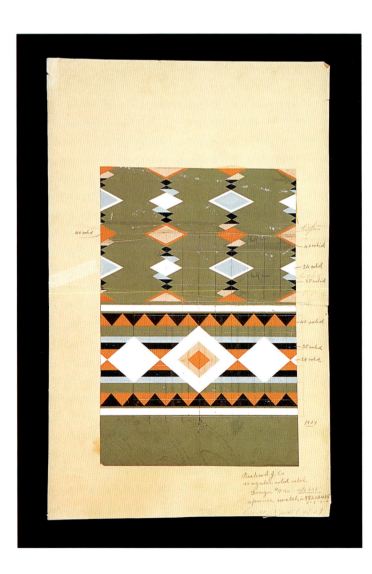

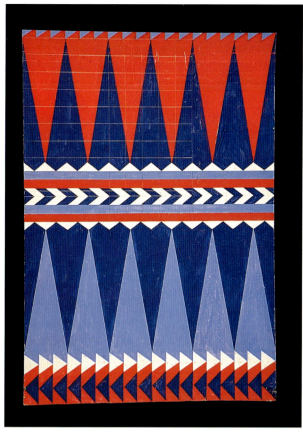

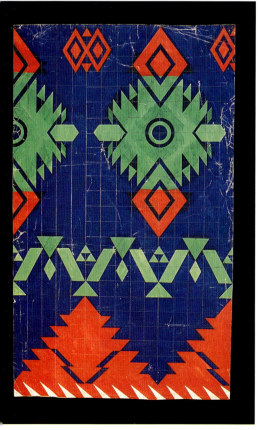

Design Concepts for Beacon Blankets

Design Concepts for Beacon Blankets continued

Chapter 3

Beacon Blankets
(Make Warm Friends)

Blanket Names & Descriptions

Listed below are most of the blanket names with dimensions and notations.

Plaid Blanket Pairs (two lengths long)

72x84", 72x90", 70x80" with 3" or 4" of sateen on the ends.
Available in all cotton, 5% to 50% wool, and as single blankets.

Usually Cotton with Plaid Designs

Midway: 60x80, ends hemmed, two color plaid with geometric border design.
York: 66x80 and 70x80, ends hemmed, two color plaids and geometric border design.
Kismet: 70x80, ends hemmed, two color plaid and geometric design border.
Mistral: 66x80, ends bound with 3" sateen, two color *Ombre* stripped with geometric border design.
Curlew: 70x80 or 72x84, ends bound with 4" sateen, two color *Ombre* plaid with geometric design border.
Yukon: 72x84, ends bound with 4" sateen, two color *Ombre* plaid with geometric design border.

Usually Cotton, Indian Designs

Toba: 60x80, economy four color Indian design blanket, not *Ombre*, with ends hemmed.
Huron: 66x80 or 70x80, ends hemmed, four color, *Ombre* Indian design, occasionally part wool.
Agawam: 60x80 or 70x80, bound all around with 2" or 3" sateen, four color, *Ombre* Indian design.
Wigwam: 70x80 or 72x84, ends bound with 4" sateen or rayon taffeta, four color, *Ombre* Indian design.
Inca: 72x84, ends bound with 4" sateen, four color, *Ombre* Indian design.

Cotton & Wool Blends

Merlin: 70x80, ends bound with 3" rayon taffeta, 25% wool with fancy borders.
Nomad: 72x84, ends bound with 4" rayon taffeta, 25% wool plaid, with geometries.

Beacon Yukon blankets, advertising painting. Courtesy of Beacon Manufacturing Co.

Roma: 72x84, ends bound with 4" rayon taffeta, 25% wool, herringbone and plaid.
Lotus: 72x84, ends bound with 4" rayon taffeta, 25% wool, solid colors with floral borders.
Casco: 70x80, ends bound with 4" rayon taffeta, 25% wool, four color *Ombre* Indian design.
Sachem: 72x84, ends bound with 4" acetate satin, 85% wool, four color *Ombre* Indian design, sold boxed.

50/50 Cotton & Wool

Belmont: 72x84, ends bound with 4" rayon taffeta, 50% cotton & 50% wool, solid color with floral boarders.

Beacon Blanket Catalogs, 1917 to 1957

The word "blanket" became part of the English language when a Flemish weaver named Thomas Blanket began producing woolen cloth that was thick and warm for use on the beds of King Edward III's household. (*Reprinted from the Beacon Blanket Manufacturing Data Publication*)

The earliest catalog in either the Beacon Manufacturing Company or the Owen Manufacturing Company archives is 1917/18. Earlier catalogs existed but have not been found.

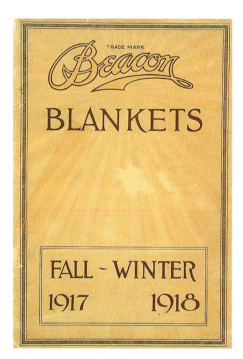

See additional pages from this catalog on pp. 48-50.

Early Beacon Blankets 1905-1925

Blankets were produced starting in 1905 and the earliest evidence of designs appearing on them (rather than being striped) is an excellent example from 1910/11. From the paper label it appears to be a salesmen's sample from that time.

Close up of the Indian Chief blanket and label. The swastika symbol means "good luck" in American Indian cultures. Courtesy of Charles D. Owen Jr. IV.

Indian Chief blanket, "c. 1910/1911." Courtesy of Charles D. Owen Jr. IV.

Patterns were carried over each year with minor variations in design and colors. Similar changes also occurred each year in the blanket programs for J.C. Penny, Montgomery Ward, and Sears, Roebuck & Co.

These subtle changes make it difficult to exactly date a blanket or shawl to a particular year. There are exceptions, however, and finding an exact match is possible. Missing catalogs may be lost or may not exist at all. With the carry-over of designs, however, approximate dates can be presumed. Design schemes were named and in most cases, but not all, fabrication remained the same.

INTEROFFICE CORRESPONDENCE
BEACON MANUFACTURING CO.

SWANNANOA   MR. C. D. OWEN, JR.   DATE   12/30/60
NEW YORK   WDM/em

Dear Charlie:

I am sending under separate cover, for your personal attention, a Wigwam blanket which was made at New Bedford in either 1910 or 1911. Thought you would like to have it for future reference.

W. D. MACRAE

The memo that accompanied the Indian Chief blanket, 12/30/60. Courtesy of Charles D. Owen Jr. IV.

A fine example of a Beacon blanket that has been to many picnics and beach outings, "c. 1910." Authors collection $125-175.

A very early example of a Beacon blanket, "c. 1911", $125-250. Author's collection.

A shocking color combination but a stunning example of an early Beacon in mint condition, "c. 1915." Courtesy of Gary Holt/Steve Christianson, Gary Holt Collection.

A subtle Indian design with an outdoors look and feeling. A blanket that has gone along on many adventures, "c. 1912." Courtesy of Charles D. Owen III.

The classic "good luck" Indian symbol, the swastika, appears as part of the design scheme on this stunning blanket, "c. 1911-1915." Courtesy of Sam and Denise Kennedy of Cisco's, Coeur d'Alene, Idaho.

Strong design elements make a powerful statement on this blanket, late-teens. Courtesy of Charles D. Owen III.

Selling a Beacon Blanket

The leaflet "Why Beacon Blankets Make Warm Friends" was prepared for sales people. It explains washing procedures and has a comparison chart of wool versus cotton blankets and an interesting explanation of the blankets' properties. The date is uncertain, but probably before 1920.

Courtesy of Bearwallow Mountain Traders, Judy Hudson & Norwood Barnes.

Why Beacon Blankets Make Warm Friends. Courtesy of Dan Owenby.

The following blankets show design patterns frequently used in the early years of Beacon, "c. 1910-1918."

Courtesy of Laura Fisher/Antique Quilts and Americana.

Courtesy of Bearwallow Mountain Traders, Judy Hudson & Norwood Barnes.

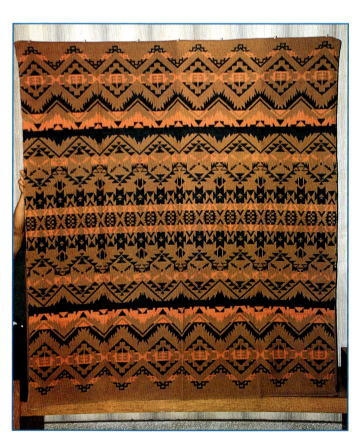

Courtesy of Sam and Denise Kennedy of Cisco's,
Coeur d'Alene, Idaho.

An outstanding example of an early Indian blanket, "c. 1919." Courtesy of Bearwallow Mountain Traders, Judy Hudson & Norwood Barnes.

The 1920s Ad Book

Beacon Manufacturing Company took a sincere interest in company-to-retailer and retailer-to-consumer relations. This publication from the 1920s (reproduced here in its entirety) shows how to sell, display, and advertise Beacon products with company-supplied sales aids, available for free. Beacon suggested, "that those interested in the sale of blankets read this advertising book so that the sales person will know how to sell Beacon blankets." Pages 21, 25, and 26 exhibit some of the art work Norman Rockwell did for Beacon. (*Courtesy Beacon Manufacturing archives and Dan Owenby*)

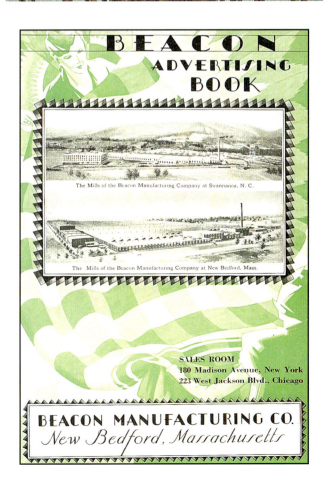

Window Display, Jas. A. Hearn and Son, Inc., New York.

Window Display, Meikles, Ltd., Bulawayo, Rhodesia, So. Africa.

One of the Beacon window cards showing a plaid blanket.

New York Salesrooms: 180 Madison Ave.

Beacon Blankets

BEACON BLANKETS MAKE WARM FRIENDS

2

INTRODUCTION

THIS thing advertising is something that interests all of us because advertising helps to move merchandise. Beacon Blankets have been featured in stores all over this country for many years. As a result of calls for advertising assistance, we have prepared this book. The new Beacon Advertising Book suggests a number of effective ways for building sales in your blanket department and in Beacon Blankets.

These suggestions have been built around your windows, your blanket department, your blanket sales person, your newspaper and other advertising. We suggest that all those interested in the sale of blankets read this advertising book so that the sales person will know how to sell Beacon Blankets.

3

4

Modern Merchandising

Success in retail merchandising depends on finding out what people want and satisfying those wants with the right merchandise. Perhaps this simple statement accounts for all the good will which Beacon blankets enjoy — the continued repeat business from old customers and the steady sales in bad times as well as good times. These retailers know that Beacon blankets are styled and manufactured with greatest attention to the requirements of the trade.

Beacon designers, through their various contacts, create new designs, and the mills work out the manufacturing details so that the new blanket is not only styled right, but constructed to meet a price in the large Beacon line.

The determining factor in styling and manufacturing is our old friend, the dear consumer. The old-time merchant made it a point to know her personally. He was there at the door to greet her as she came into his store. He frequently shook hands with Mrs. Jones and Mrs. Brown and patted little Johnny on the head and inquired about Grandma. He knew many of his customers by their first names. He knew their buying habits. He knew where they lived and what they bought.

Today he is the head of a complicated retail organization. He can no longer personally greet or wait on Mrs. Jones, and Mrs. Jones has changed her ideas on blankets. Styles in patterns and colors were unimportant. Today they are the difference between a popular line and a slow-moving line.

The modern woman buys as her fancy dictates. She doesn't pop into the first store and buy the first blanket she sees. As purchasing agent of her home, she figures out certain things, perhaps not always in the same order, but still she figures them out.

The blankets are for a certain purpose; that is, for a double bed, single bed, crib, carriage, automobile, camp, for the beach, and so on. Then she thinks in terms of colors. The room is perhaps her bedroom or the guest room. Because she is a modern up-to-date woman, she has a color scheme. During the past few years she has become more and more color-conscious. She has seen an attractive bedroom at the home of a friend, or she has been reading some of the many women's magazines on home decoration. So that today when she thinks of blankets, she remembers the color scheme of the room in which the blankets are to be used.

If she likes orchid, she may have orchid curtains at the window, an orchid bedspread on the bed, and even orchid sheets and pillow slips. She will probably like to have orchid blankets in solid color, block design or plaids in which orchid predominates.

To meet this need you must have orchid blankets, and because Beacon Blankets are the best of their grade, you will need orchid blankets. Beacon blankets are made in a number of different styles, all illustrated in full color in the Beacon catalog.

If you haven't a representative line, go through the Beacon catalog and select the types of blankets which fit the needs of the type of people who trade at your store.

After you have bought the right merchandise, the next step is to expose it for sale. Mrs. Jones wants to see what she wants to buy before she buys it. And before she goes to see it, she wants to know where she can see it, and know other interesting blanket facts.

So you have to advertise — you have to tell people in print who, what, where, when, and how much so as to get your share of the retail business. To many dealers the display window is the first means of reaching out for customers. In the next few pages we will explain Beacon advertising helps for window displays.

5

---

cause people to stop and look, but also cause them to enter the store and purchase merchandise displayed.

You must plan in advance for your window display of blankets. There are many things that suggest ideas for good displays. You should plan displays ahead and get everything needed together before starting in to work in the window. Your window must advertise the type and character of your store. Try to put human interest into the displays, suggesting the use of the articles shown. In general, all relative articles should be shown together. Study other store displays. Merchants in other lines of business have good ideas that you can adapt to your own use. You will find it valuable to keep a notebook record of ideas you may get from other stores' windows and trade papers.

Another point to remember is not to crowd your windows. Also avoid the other extreme of too little merchandise. Help the eye to travel to related subjects by running a tape or ribbon from cards to the blankets and displaying the price on the merchandise shown. Strive for a balanced window; that is, determine on a central unit such as a large Beacon cutout. On one side put a pile

Window Display, Steiger-Dudgeon Co., New Bedford, Mass.

7

Shopping Your Windows

Window Display by the Home Store, Dayton, Ohio. With a Turn Table in the center to give continuous action.

Mrs. So-and-So is coming downtown today to buy some ribbon. She passes your window. Does she stop, look, and become interested? Her actions depend on the attractiveness of your display which causes her to become interested in your window, and the effectiveness of your display which holds her attention until she decides, "Yes, I want that," or "No, I don't like that."

Window decorating presents a group of problems as varying and as important as any connected with retailing. The window functions as a visual salesman. It attracts the customer, arouses interest, and makes the customer enter the store to find out more about those articles displayed.

Each window has its own points, and better results may be expected from one window than another. It may be on a street where many pass or very few. It may have good or poor light. All these things are important.

Lighting effects and background should be arranged to give the best possible display of merchandise. While you may have a Beacon blanket window today and next week something very different, your window displays should be treated in the best possible taste for the particular articles shown.

The sidewalk view of your window is an assemblage of detail and ensemble to which the window frame, the background, the lighting, and other things are just as important as they would be for a presentation on a stage. Study your window and see if it is too deep for good visibility or too high to focus attention upon the merchandise display. Corrections should be made so that you will get the best effectiveness.

The window is to sell goods. If it does not do this, it is a failure. It must not only

6

Window No. 1

Window No. 4

Window No. 2

Window No. 5

Window No. 3

Window No. 6

Simple unit displays for small windows or units in larger windows
Each display centers around a Beacon cutout or card

8

37

of blankets which can be balanced on the other side by a rack with a blanket thrown over it.

Your lighting should be the best possible. Block off the store interior so that it can not be seen from outside the window. The back of the display or the partition should be high enough to shut off all view of the interior. A dark-colored background, when displaying light merchandise, will emphasize the contrast of colors. The windows should be spotlessly clean and washed daily if possible. The display itself must be spick-and-span. Never allow soiled merchandise in the windows.

To help make good Beacon blanket windows with least trouble, we have prepared what we term practical window units using

Beacon advertising displays as a basis.

The first window is a "Boy Display" based on cutout No. 703.

The second window is a "Modernistic Window" based on cutouts Nos. 738 and 739.

The third window shows twin beds set up without the side boards. This arrangement gives the bedroom atmosphere to the small window. See Window Card No. 700.

The fourth window is the "Girl Display" based on cutout No. 733.

The fifth window is another human interest boy window. This window is built around the Indian patterned blankets. Cutout No. 702.

The sixth window is the "Beacon Crib Blanket Display" and is based on cutout No. 711.

Window No. 7, showing large cutout and two small cards.

9

Window Card No. 225

Window Card No. 221

Window Card No. 219

Window Card No. 226

Window Card No. 211

Window Card No. 227

Beacon Window Cards reproduced in full color. Size 22" x 28"

The seventh window is called "Lady Display." It is built around cutout No. 732 — a life-size lady holding a Beacon blanket. Note the simple arrangement with only a comparatively few blankets in the window.

These seven displays are easily set up. They are primarily suggestions for units, but may be used as windows without the addition of more merchandise. Of course you can easily build much more elaborate windows. Whatever you do, we would like to have photographs of your Beacon windows. If you will have two photographs made, keep one and send the other to us together with the photographer's bill. We will gladly pay for both sets of photographs.

11

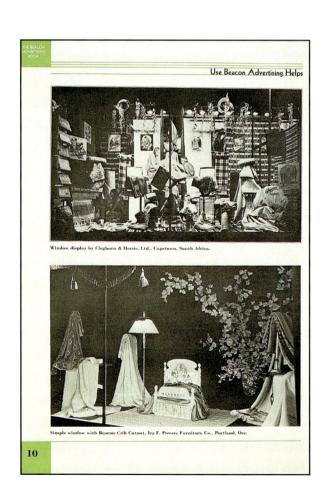

Window display by Cleghorn & Harris, Ltd., Capetown, South Africa.

Simple window with Beacon Crib Cutout, Ira F. Powers Furniture Co., Portland, Ore.

10

Window Card No. 229

Window Card No. 232

Window Card No. 234

Window Card No. 230

Window Card No. 233

Window Card No. 235

Window Card No. 231

Window Card No. 236

BEACON WINDOW CARDS reproduced in full color Size 22" x 28"

12

38

Beacon Cutouts & Window Cards

All Beacon cards are colorful and in themselves attractive suggesting different uses, different patterns, and different color combinations. They range in general art treatment from the most conservative to the modern. The cards vary in technique, including watercolor, oil, pastel, and charcoal. Some are printed and some are lithographed. Thus we have endeavored to give you cards which appeal to you personally as well as cards which will help you sell merchandise. One-color plates, electros of the cuts shown in this book, are available for any dealer to use in his direct-mail advertising. A few of these cards are also shown on page 46. If you wish any of those shown on page 11 or 12, please write to us specially indicating what illustrations you would like to have. We suggest their use in mail-order effort.

This year we have prepared a number of new general cutouts and window cards. New ones are being brought out periodically. Those now ready for distribution are shown in this book. If you haven't any of them, write to the Advertising Department, New Bedford, for additional display material for use in your store windows, in your blanket department and in booths at fairs and other special occasions.

Cutout No. 738

This page shows a pair of modernistic window cutouts — most attractive and helpful in building a pleasing window.

Cutout No. 739

13

Direct-Mail Advertising

The question is — how to reach customers and prospects whose names and addresses you have or can get? The answer is — direct-mail advertising.

Much of the direct advertising is being written by an individual who has many other jobs. He dashes off a form letter and thinks it will cover the direct-mail advertising as far as he is concerned. It is usually not written carefully or with any careful plan. It is full of "we's." It starts in about the latest merchandising triumph instead of thoughtfully helping the reader to visualize new blankets in her home.

Direct mail is an ideal method for reaching people — such as charge customers, families in the telephone book, club members, and so on, because you have their names and addresses ready at hand.

The rules to follow when preparing direct advertising on Beacon blankets are, first, as in other forms of advertising, be sure to have the merchandise. Second, determine how much you can spend on the direct-advertising piece. Then prepare the text and see about illustrations. Set up your mailing list, and see that the piece is printed, addressed and mailed on time. After you have written the text lay it one side and do something else. Later read it through again. Does it

sound good? Is it newsy? Will it interest somebody? Will it make that person come in and see your blankets? Is it properly illustrated?

Make any changes and corrections you want before you send it to your printer. Try to make as few as you can when the proof is returned.

Post a copy of all your advertising in the department concerned. Instruct your sales people about the merchandise and have it well displayed.

Make use of the different Beacon cuts shown in this book — line plates or halftones. They will make your mail-order effort much more effective.

When customers ask for Beacon blankets you do not carry, let them order from the Beacon catalog. Keep one handy in your blanket section. If you wish another, we will be glad to send you one. If there are any blankets with which you are unfamiliar, we will be glad to send you a sample blanket.

To save you the trouble and much expense in preparing direct advertising, we have produced some Beacon folders which you can use. The folders are attractive consumer advertising showing reproductions of the blankets in full color.

These folders may be sent through the mail

Window Display, Wolf's Quality Shop, Shelbyville, Ind.

15

Use Beacon Advertising Helps

Cutout No. 734, Plaid Blanket.

Cutout No. 735, Indian Blanket.

Make Warm Friends

Cutout No. 733, Plaid Blanket (center piece).

ATTRACTIVE
LARGE WINDOW CARD
WITH SMALLER SIDE
CARDS

Cutout No. 736, Jacquard Blanket.

Cutout No. 737, Crib Blanket.

Beacon Blankets
make Warm Friends

Window Card No. 700. Size 36" x 45".

14

New York Salesrooms: 180 Madison Ave.

to your customers or used as package enclosures. We will send you a supply of folders if you will tell us how many you can use.

If you are going to send out a general mailing piece to your customers, you will want to get proper blanket representation. In the back of this ad book you will find attractive halftone illustrations of Beacon blankets for such use. On page 46 are six special halftone reproductions of window cards in a size suitable for use in direct advertising. They may be used in the preparation of counter folders or package inserts, or in a number of different ways. Direct advertising via the government post card is perhaps the least expensive, but good results are obtained by using folders, cards, etc., or envelope en-

closures or distributing them by boys from house to house. This latter method is an inexpensive way of reaching people, but it is also apt to be wasteful when it comes to letters.

Remember that you must get a feminine appeal into them. Perhaps you may wish to mention price, and as a rule prices appeal to a woman. But she also craves beauty and social prestige. New blankets in keeping with the color scheme of her room give her something to think about and talk about. This is good advertising for you provided to help educate her and get her to buy the blankets from you and not your competitor. Read over these direct-advertising suggestions the next time you start to prepare a folder or mailing card.

Window Display at the I. P. Gerhart Store, Clarksville, Tenn.

16

39

Blanket Descriptions
FOR USE IN WRITING ADVERTISEMENTS

The following paragraphs give a brief description of the different types of Beacon blankets shown in our catalog. If you are unfamiliar with any of these blankets, we will send samples. Use these blanket descriptions as a basis for your copy. For general information on construction, washing, etc., read pages 21 to 26.

H- and P-Plaid Blankets

Beacon H- and P-Plaid blankets are a necessity in every home. Even in wintry weather, their exceptional warmth and generous tuck-in afford ample protection. The newest patterns are more pleasing than ever, including unusual three-color effect. Size 70 x 80. Pairs. Ends bound with 4-inch sateen with four rows of stitching.

B Blankets

Beacon B Blankets are made in six solid colors with a two-bar border. Size 70 x 80. Pairs. Ends bound with 4-inch sateen with four rows of stitching.

X Blankets

These single X Blankets are in solid color with three-bar border in contrasting colors. Size 70 x 80, ends bound with 3-inch sateen.

Window by R. A. DeFreest, Cohoes, N. Y.

Priscilla Blankets

Beacon Priscilla Blankets present conservative borders with the plain solid-color back-

Window by R. A. DeFreest, Cohoes, N. Y.

grounds and solid-color reversible pastel shades. Blue, rose, orchid, green, gold, and tan. Size 66 x 80. Single. Ends bound with 3-inch sateen.

Signet Blankets

Beacon Signet Blankets offer plaid checked decorative effects that are new, and exclusively Beacon. They are adapted to scores of uses. Their rich colorings make them particularly suited to replace the spread. Size 66 x 80. Single. Ends bound with sateen.

Plaid Blankets (single)

Single Beacon Plaid Blankets are made in two weights — medium and heavy — and in three sizes — 66 x 80, 70 x 80, and 72 x 84. All blankets are unusually attractive in pattern and color. Each bound on ends with both 3-inch and 4-inch sateen.

P Y Blankets

Beacon P Y Blankets are another line, showing a solid-color plain-bodied blanket with Ombré border. Size 70 x 80. Single.

17

Use Beacon Advertising Helps

Window Display by Steiger-Dudgeon Co., New Bedford, Mass.

Bound on ends with 4-inch sateen with four rows of stitching. Colors blue, rose, orchid, green, gold, tan with contrasting Ombré border effects. Beautiful blankets for any room.

P-C Comfortables

Beacon P-C Jacquard Comfortables serve as additional covering at night. Daytimes they take the place of the bedspread. The patterns feature unusual three-tone effects. Sizes 72 x 84. Single. Bound around with 3-inch superior quality satin ribbon.

P R Blankets

Beacon P R Blankets are reversible solid-color blankets with blue one side, maize the other, rose and maize, rose and blue, peach and maize, orchid and peach, green and rose, blue and rose, green and orchid, peach and green. Bound all around with 4-inch sateen with four rows of stitching.

Size 72 x 84. Single. Ideal extra blankets. Warm and colorful. Will tone in with color schemes of any room.

Ombré Blankets

Beacon Ombré Blankets are heavy top blankets. They come in fine rich colors and are ideal to use as extra blankets or throws. Size 70 x 80. Single. Ends bound with sateen.

Indian Design Blankets

Beacon Blankets, — a colorful addition to any couch, bed, porch hammock, or Cape Ann cot. Because of their warmth, convenience, and utility, they are popular also for motoring, boating, canoeing, and outdoor purposes. Bound all around with sateen. Size 60 x 80. Single.

Ombré Plaid Robes

Beacon Ombré Plaid Robes are new and very desirable for auto robes and camp-

18

New York Salesrooms: 180 Madison Ave.

ing. They are in the new bright colors of Ombré plaid. Single. Size 60 x 80. Bound all around with sateen.

Bathrobe Blankets

Beacon Bathrobe Blankets offer Beacon quality in colorings and patterns that are truly wonderful. Each blanket is large enough to make a full-size garment, and we have girdles to match.

Bathrobe Cloth

Beacon Bathrobe Cloth (with girdles) enables you to purchase the required amount of material in 27-inch or 36-inch width. Patterns include Ombré effects, also nursery designs for children's robes.

Crib Blankets

Beacon Crib Blankets come in several sizes affording ample tuck-in for crib, bassinet, carriage, or porch basket. Made to stand

frequent washing without impairing their warmth-retaining qualities. The newest are the blue, pink, rose, orchid, green, and maize pastel shades in plain reversible blankets. Bound and stitched. Sizes 30 x 36, 30 x 40, and 36 x 50.

Child's Bed Blankets

Beacon Child's Bed Blankets come in pink and blue (solid color), blue and white, and pink and white (block), size 42 x 66, bound around with 3-inch sateen. These blankets are popular with mothers who want a generous-sized blanket to go on the bed of the child who has outgrown her crib.

Fringed Carriage Robes (crib size)

These Beacon robes are used when baby is taken out for an airing, or an emergency extra bed blanket. They are made in two sizes, medium and large, — fringed all around. The colors include blue and pink with fancy jacquard borders, solid colors, and plaids.

Window display at the L. E. and H. J. Hamilton Store, Albany, Oregon.

19

Use Beacon Advertising Helps

Beacon Blankets make Warm Friends

In shades to blend with your color scheme

Window card No. 236 reproduced in full color. Size 22" x 28".

20

Talking With Your Customer

Note. The following comments and suggested sales talks are intended for new sales people who have just started in to sell over the counter. While we are naturally thinking in terms of blankets, the same basic method of handling the sales talk is applicable in selling other merchandise. We suggest that you allow us to send additional copies of this booklet to your newer sales people.

Place: Blanket Town Department Store.
Characters: Two sales girls and a customer.
Scene: Blanket counter.
Time: Any time.

1st Sales Girl: What's new, Mabel. Been to a show lately?

2nd Sales Girl: Yeah, last night. But it wasn't much. Say, did Mamie tell you about the chap she went out with last —

Customer: Pardon me — but —

2nd Sales Girl: — night. A big brief case and sample man selling cloaks. No—o?

1st Sales Girl: Naw — she did n't breathe a word. What happened?

Customer: Excuse me, but how much are these blankets?

1st Sales Girl: [Aside] Wait a minute. [To customer] What's that, Madam?

Customer: I want an orchid blanket. How much —

1st Sales Girl: Have n't orchid, Madam, sorry — [To 2d Sales Girl] Go on, what about it?

2nd Sales Girl: Well. She took him up an' —

Customer: Are these all the blankets you have?

1st Sales Girl: We haven't any orchid, Madam. The nearest we have is lavender.

Customer: Let me see that. Perhaps that will do.

1st Sales Girl: [While pulling

orchid or lavender blanket from pile — to 2nd Sales Girl] Where'd they go? [Pile falls over — sales girl throws orchid blanket on counter and says to customer] There y'arc. [To 2nd sales girl] Yeah?

Cutout No. 711. For featuring crib blankets. Size about 31" x 40"

Customer: Is that all you have in orchid?

1st Sales Girl: That's lavender and that is all we have. Sorry.

Customer: Well — never mind.

1st Sales Girl [To 2nd Sales Girl] Don't some women drive you to drink. That's lavender not orchid. Well, tell me what happened.

* * *

This little scene is an exaggeration of the sales methods in most stores. Nevertheless, it brings out certain principles of retail salesmanship. The basis of successful retail selling is confidence, education, desire, and volition. You gain the customer's confidence in you. You will frequently have to educate and create on her part a desire not only to trade with you, but to trade always with your store. Finally you must learn how to

close the sale—help the customer decide to buy what you have to sell.

Confidence building starts with the moment a customer enters your department. When she hovers around your counter, she is probably there because she wants information about blankets. She may have come in because she wants to buy a blanket. She may be in the store to purchase some other article and is attracted to your blankets by a special display or sale, or she may be waiting for a friend. Whatever the reason, she has come there of her own volition.

What should you do and say when the new customer comes in? You should discontinue any conversation with other employees. "Size up" your customer and ask her a definite question *"May I show you a Beacon blanket?"* Your first words are important. You want to get her thinking about

Window display at Holzeimer & Shaul, Amsterdam, N. Y.

---

something definite and tangible — not a blanket, but a particular blanket — and want to get her agreeing with you. By asking her if you may show her a Beacon blanket, you have expressed a definite idea and aroused a mental picture of a type of blanket with which she is already familiar. This is a starting point. She must answer, *"Yes, I want so-and-so"* or *"No, I want so-and-so."* Your approach to her and the manner of her reply has given you time to study her. She may be in a hurry or have plenty of time. She may be poor or wealthy, old or young, married or unmarried, quiet or gay, definite or indefinite, good-natured or cross and tired.

Suppose that she is apparently a woman of some means. She may care nothing for price, but on the other hand, she may be very thrifty. First interest her in the high quality and beauty of Beacon blankets. It is easier to trade her down than up. Find out her favorite colors and for what use she wants the blanket. Tell her of the durability, of the excellent workmanship, of the variety to choose from, of the refinement of color and patterns.

If she is undecided just what she wants, try first to learn the situation, whether it is for a gift and for whom, or for herself and if so, her color scheme. Make suggestions to her, but be careful not to give so many that you make her even more undecided. Be patient and tactful with a rich woman, because she can afford to go where the price is higher if she does not like what you are showing her.

Now let us turn to the poor woman. She is less apt to be undecided, because she knows what she wants, a warm low-priced blanket. Usually color means little to her except that she likes things "pretty." After calling her attention to the fact that the blankets are "pretty," show her how warm they are, how

Cutout No. 732. Life-size lady holding Beacon Blanket. A real blanket can be draped as desired. Size about 25" x 63".

long they will last, and then if she seems interested, talk color harmony.

Most of the women before the counter are neither rich nor poor, though individually they may have personalities like a rich one or a poor one. In general they are practical, and it is well to stress the quality. Tell them Beacon blankets are the best made at that price, and of a higher quality than many

---

MADE IN NEW BEDFORD

Window display by Steiger-Dudgeon Co., New Bedford, Mass.

others at the same price. The women are especially interested in interior decoration, and do much experimenting along those lines, often doing all their work themselves. They have definite color schemes. It is well to find out first of all just what their tastes are, and try to picture for yourself what the home of each looks like. If the customer is hazy about what would fit her scheme of decoration, make suggestions, bearing always in mind that she is always critical and hence your suggestions must be sensible and very profitable. Convince her that Beacon blankets are the best looking, the most durable, and the warmest for price anywhere, and those who can afford higher-priced blankets prefer to buy Beacon blankets.

Always study the woman herself and be awake for little hints that may tell you on

what grounds she may be appealed to, and consequently be made to buy. Regardless of who or what she is, her answer to the question, *"May I show you a Beacon blanket?"* will immediately tell you what to say next, and how to say it.

If she replies, *"I am looking for something for my guest room. Not too expensive,"* you know something of her needs. But you want to know more. So, as you lead her to a display of Beacon blankets, you ask for more information, such as what color had she in mind.

When you remind her that people are buying blankets to go with the color scheme of the room, you are appealing to her intelligence and demonstrating your own knowledge, thus giving confidence. The fact that draperies, wall paper, rugs, pictures, bed-

Order Beacon Cards and Cutouts by Number

spreads, sheets, pillow slips, blankets, and even the woodwork and the furniture itself are related in color is interesting, if nothing else.

You need not go further into the ensemble idea at this time, but turn to a pile of Beacon blankets and pick out one in the color she mentioned and ask her to feel of it. Now you have turned from the general to the specific. You are progressing and you say, pleasantly, *"Do you like this Beacon blanket? It is our H-plaid. Part wool, size 66 x 80."* Point to the Beacon ticket which gives size, etc. Now you are directing the sale, yet taking advantage of every opportunity.

If her answer is a firm "yes," registering a marked degree of interest in that particular blanket, you have probably struck fire and are on the way to close your sale. Let us suppose she only agreed with you that it was a nice blanket. You haven't struck fire, so you must go on at once in an endeavor to stimulate interest.

Display a knowledge of your merchandise. Get her confidence by telling her blanket facts which ought to interest her. *"These Beacon blankets are warmer than many all-wool blankets."* If she makes no further comment, lead the way to another pile of blankets and find

out just what will strike fire. If she counters *"Is that so?"* even registering only a little interest, follow up with Beacon facts which will interest her.

Pick up the blanket and continue holding the customer's attention. *"I'll show you why. Beacon blankets are part cotton and part wool, madam. The cotton comes from China, India, and the South, selected for grades most suitable for blanket purposes. The cotton is processed through machinery similar to that used in preparing and cleaning wool. It is further carded and spun on the wool system. Wool threads are so mixed with cotton that with the wool process, a woolly effect is produced. You see, madam, that Beacon blankets are constructed so as to gain best quality for a given weight blanket. The Beacon people do this without regard for quantity production. This makes Beacons, grade for grade, warmer than any other part-wool blankets on the market and warmer than many all-wool blankets. Now, let me prove it.*

"Just feel that blanket, madam. Isn't it fluffy and soft? That is what we call loft. A Beacon isn't thin and hard like a sheet. It has thickness. Well, that thickness — all those little ends — hold still air, which

Cutout No. 702. Size 38″ x 41″. Place order by number.

25

The Newspapers

"What'll we put in the newspaper ad tomorrow?" Examine the advertisements and cuts shown in this book and you will have the basis for a good newspaper ad.

Successful writers of retail advertisements tell us that advertising copy involves, among many things, these three features: (1) that you have something to say about your blanket story; (2) that you say it interestingly, newsily, and in plain English which anybody can understand; (3) that you put your blanket story where people will see it.

Feature in your advertisements standard salable merchandise, such as Beacon blankets. Don't use the newspaper just for price cutting and sale events. Feature regular merchandise except at sale times.

In preparing the advertisement, illustrate it properly. Use Beacon illustrations. Get the "you" attitude, not the "we." Put news and salesmanship into the text. Have a good caption or display heading. Don't crowd your advertisement just to save

Beacon Blankets

make Warm Friends

For the Particular Hostess

■ We recommend Beacon blankets because Beacons are styled most attractively with blending or contrasting colors. You will find your guests appreciate Beacon blankets.

■ Let us show you our assortment of blankets. Double Blankets from $0.00. to $0.00

THE VOLK STORE
MERIDAN STREET

27

New York Salesrooms: 180 Madison Ave.

is a non-conductor of heat. That's how Beacon construction and finish give such a soft, warm blanket. And then, too, the close weave makes Beacon blankets wear well in spite of many washings."

Your customer will probably say: *"I see, I never thought of that before."* You have given her new ideas about blankets — information which she will take pride in telling to somebody else. She becomes more interested and asks — *"What have you in rose?"*

This is your opportunity to assist in guiding the color selection. You ask her, *"Is rose to be the predominating color in your room?"* You will want to remind her again of the ensemble idea used by leading decorators. Continue by saying that color schemes for bedrooms usually start with the draperies,

and that the blankets should harmonize with the curtain color scheme.

You can explain that all Beacons are fast colors, and that all dyeing is done in raw stock with colors fast to light and washing.

You must clear away any objection about washing by stating that this construction prevents any fading or lumping, and that by brushing the blanket with a stiff brush, she can make the finish look like new. Also tell her that any high-grade laundry will wash and renap her blanket in a special blanket carding machine designed especially for laundries.

Get the customer to feel each blanket of a half-dozen patterns. Finish is something which has to be felt.

The hardest part of the sale comes now, getting her to say, *"I'll take that one."* To her question as to price, tell her something more than the figures, say that *"This Beacon pastel blanket in the color you like is . . . . . dollars. It is one of the highest-grade blankets manufactured by the Beacon Company."* Thus, you haven't left the price hanging in mid-air. You have followed up with facts which will make her believe the blanket is worth the price. Then continue: *"From what you have said, I think this blanket will fit in with your color scheme and will look well in your guest room. I know it will wear well and keep your guest warm, and those are three big factors, aren't they?"*

In this way you are making it hard for her to put off buying. You are forcing her to agree with you and to put herself in the frame of mind for immediate buying. All she needs now is the final urge if she hasn't already agreed to the sale. So you say, *"Shall I have it charged and send it out?"* or *"Shall I have this blanket sent to your home?"* and your sale is made.

Cutout No. 703. Size about 40″ x 48″.

26

Use Beacon Advertising Helps

Beacon BLANKETS

make Warm Friends

● Now that cold weather is approaching it is time to think of fresh blankets in colors and patterns which will harmonize with your bed rooms.

● A new stock of these soft, warm Beacon blankets has just arrived from the mills. You will find them in our blanket department.

● The new Beacon pastels, plaids, and blocks are most attractive. Single Blankets $0.00 to $0.00. Double Blankets $0.00 to $0.00.

BOSTON STORE
HOUSE FURNISHINGS — THIRD FLOOR

28

a line, because a crowded advertisement is hard to read and therefore, read less than an advertisement with a little white space to set it off. Be truthful in all advertisements, for in no other way will you keep the consumer's confidence.

When deciding whether or not you will advertise in the newspapers ask yourself if the people who read the pa-

BLANKETS MAKE WARM FRIENDS

When Style and Value Meet

■ Today the modern housewife considers color as well as quality when shopping for blankets. She wants style as well as value.

■ Beacon blankets give you not only the modern stylings in color and pattern, but also the quality which comes as the result of many years at manufacturing blankets.

■ We have a large assortment of Beacon and other high-grade blankets from $0.00 to $0.00 per pair.

For the Baby

■ He's such a cute little fellow — so small and dependent.

■ See that he has plenty of soft, warm Beacon Crib Blankets . . . $0.00 to $0.00.

TOBEY'S
27 Main Street

The Arcade
2ND FLOOR
288 TRAVISTOCK STREET WEST

29

Beacon Mats and Electros
PLEASE ORDER BY NUMBER

No. 3001-A, one column (2" x 1½")
No. 3001-B, two columns (4" x 3½")

No. 3002-A, one column (2" x 1½")
No. 3002-B, two columns (4" x 3")

Beacon Blankets
make Warm Friends

No. 3003-A, one column (2" x 1½")
No. 3003-B, two columns (4" x 2¾")
No. 3003-C, three columns (6" x 3⅝")

31

MODERNISTIC COLORS

Beacon
BLANKETS
make Warm Friends

■ How soft and luxurious these Beacon blankets feel—yet how inexpensive!

■ Perhaps that is the reason for their great popularity.

■ Before you make any blanket purchases see our list of Beacon's plaids, blocks, and solid color blankets— Single blankets $0.00 to $0.00 Double blankets $0.00 to $0.00.

Starr Brothers

pers are the people who would come to your store to buy blankets. Then decide whether the cost of the newspaper space needed to tell your story properly is commensurate with the business you can get from the advertisement. If there is more than one paper in your city, which paper should you use? On which page do you want your advertisement to appear? Space is purchased by the column, inch, or line. Perhaps your paper gives a lower rate to certain businesses, or gives a special discount under certain conditions. Talk with your newspaper man and see that you get the best possible rate. Give him the text and cuts or mats. Get him to set up your advertisement so that it will look like one of the sample advertisements shown here. Then tell him to set up all advertisements in this general style. Have him show proof to you and "O K" it before the advertisement is run. Vary your cuts and text.

On the following pages are shown Beacon newspaper illustrations. They are made up in one-, two- and three-column widths as indicated, under each illustration. Order mats or electrotypes from us.

30

32

43

Page 33

Please Order These Mats and Electros by Number

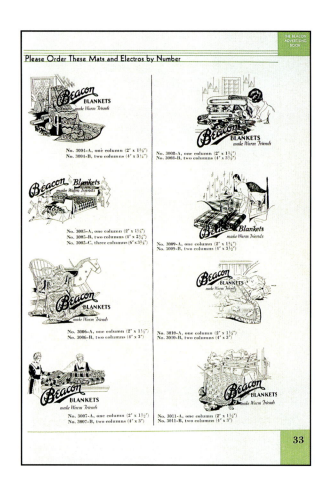

No. 3004-A, one column (2" x 1¾")
No. 3004-B, two columns (4" x 3¼")

No. 3008-A, one column (2" x 1¾")
No. 3008-B, two columns (4" x 3½")

No. 3005-A, one column (2" x 1¾")
No. 3005-B, two columns (4" x 2¾")
No. 3005-C, three columns (6" x 3⅝")

No. 3009-A, one column (2" x 1¾")
No. 3009-B, two columns (4" x 3½")

No. 3006-A, one column (2" x 1½")
No. 3006-B, two columns (4" x 3")

No. 3010-A, one column (2" x 1½")
No. 3010-B, two columns (4" x 3")

No. 3007-A, one column (2" x 1½")
No. 3007-B, two columns (4" x 3")

No. 3011-A, one column (2" x 1½")
No. 3011-B, two columns (4" x 3")

33

Page 35

Use Beacon Advertising Helps

No. 3015-A, one column (2" x 2")
No. 3015-B, two columns (4" x 4")

No. 3019-A, one column (2" x 2")
No. 3019-B, two columns (2" x 3")

No. 3016-A, one column (2" x 1¾")
No. 3016-B, two columns (4" x 3½")

No. 3020-A, one column (2" x 1½")
No. 3020-B, two columns (4" x 3")

No. 3017-A, one column (2" x 1¾")
No. 3017-B, two columns (4" x 3¼")

No. 3021-A, one column (2" x 2¼")
No. 3021-B, two columns (4" x 4½")

No. 3018-A, one column (2" x 1¾")
No. 3018-B, two columns (4" x 3¼")

No. 3022-A, one column (2" x 1½")
No. 3022-B, two columns (4" x 3")
No. 3022-C, three columns (6" x 4½")

35

Page 34

New York Salesrooms: 180 Madison Ave.

No. 3012-A, one column (2" x 1½")
No. 3012-B, two columns (2" x 3")

No. 3013-A, one column (2" x 1⅛")
No. 3013-B, two columns (4" x 2¼")
No. 3013-C, three columns (6" x 3¾")

No. 3014-A, one column (2" x 1½")
No. 3014-B, two columns (4" x 3")

34

Page 36

36

44

No. 3023-A, one column (2" x 1⅜")
No. 3023-B, two columns (4" x 2¾")

No. 3024-A, one column (2" x 1⅝")
No. 3024-B, two columns (4" x 3¼")
No. 3024-C, three columns (6" x 4⅝")

No. 3025-A, two columns (4" x 1½")
No. 3025-B, three columns (6" x 2¼")

37

No. 3031-A, two columns (4" x 3¼")
No. 3031-B, three columns (6" x 4⅞")

No. 3032-A, two columns (4" x 2")
No. 3032-B, three columns (6" x 3")

No. 3033-A, one column (2" x 1¾")
No. 3033-B, two columns (4" x 2½")
No. 3033-C, three columns (6" x 3¾")

39

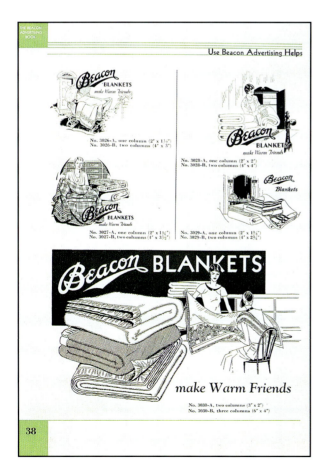

No. 3026-A, one column (2" x 1½")
No. 3026-B, two columns (4" x 3")

No. 3027-A, one column (2" x 1¾")
No. 3027-B, two columns (4" x 3½")

No. 3028-A, one column (2" x 2")
No. 3028-B, two columns (4" x 4")

No. 3029-A, one column (2" x 1⅜")
No. 3029-B, two columns (4" x 2¾")

No. 3030-A, two columns (3" x 2")
No. 3030-B, three columns (6" x 4")

38

40

45

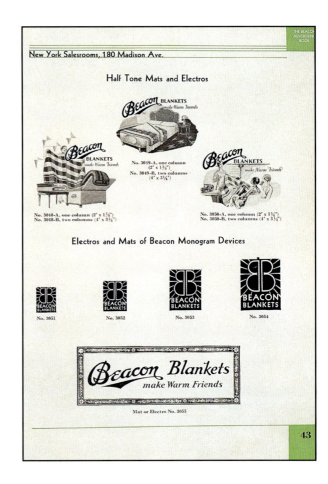

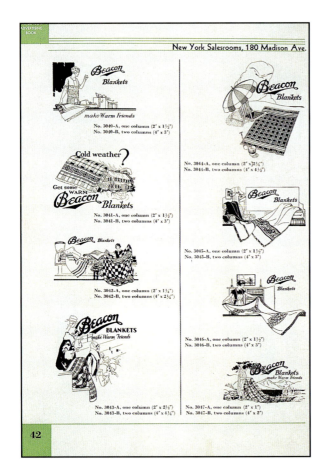

46

Direct Advertising Electros

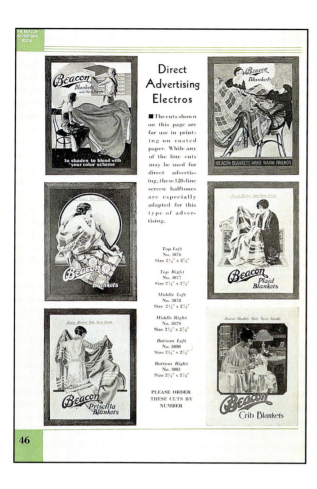

■ The cuts shown on this page are for use in printing on coated paper. While any of the line cuts may be used for direct advertising, these 120-line screen halftones are especially adapted for this type of advertising.

Top Left
No. 3076
Size 2¼" x 2⅞"

Top Right
No. 3077
Size 2¼" x 2⅞"

Middle Left
No. 3078
Size 2¼" x 2⅞"

Middle Right
No. 3079
Size 2¼" x 2⅞"

Bottom Left
No. 3080
Size 2¼" x 2⅞"

Bottom Right
No. 3081
Size 2¼" x 2⅞"

PLEASE ORDER THESE CUTS BY NUMBER

46

TABLE OF CONTENTS

Use Attached Return Cards to Order Beacon Advertising Helps

Written and printed by the Tolman·University Press, Cambridge, Mass., U.S.A.

Send for this Free Advertising Material

■ The following advertising material is furnished Beacon retail dealers free of charge. To secure any of these advertising helps fill out and mail one of the attached post cards. Write plainly your needs. Material sent charges prepaid.

No.	Description	Page
211	Plain Blankets	11
219	Yukon Blankets	11
221	Robe Cloth	11
225	Signet Blankets	11
226	Priscilla Blankets	11
227	Plaid Blankets	11
229	Plaid Blankets	12
230	Mistral Blankets	12
231	Priscilla Blankets	12
232	Plaid Blankets	12
233	Plaid Blankets	12
234	Plaid Blankets	12
235	Crib Blankets	12
236	Reversible Blankets	12
700	Beacon Blankets	14
702	Boy Asleep	25
703	Boy in Chair	26
711	Girl and Puppies	21
733	Plaid Blankets	14
734	Plaid Blankets	14
735	Indian Blankets	14
736	Jacquard Blankets	14
737	Crib Blankets	14
738	Beacon Blankets	13
739	Beacon Blankets	13
3001–3065	Newspaper Line Cuts	31–44
3066–3075	Newspaper Halftones	45
3076–3085	Halftones	46–47

48

New York Salesrooms, 180 Madison Ave.

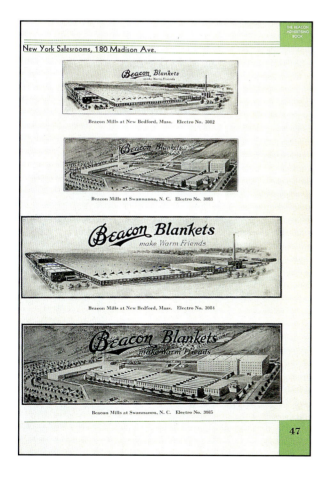

Beacon Mills at New Bedford, Mass.   Electro No. 3082

Beacon Mills at Swannanoa, N. C.   Electro No. 3083

Beacon Mills at New Bedford, Mass.   Electro No. 3084

Beacon Mills at Swannanoa, N. C.   Electro No. 3085

47

SAVE THIS Beacon AD BOOK

Beacon PLAID AND FANCY BLANKETS

STYLE P 52

STYLE P 55

STYLE P 82

STYLE P 1

Above are two-color designs.   *For colors and descriptions see page 31.*

BEACON MANUFACTURING CO., PROVIDENCE, R. I.
New York Salesrooms, 50 Union Square East.
29

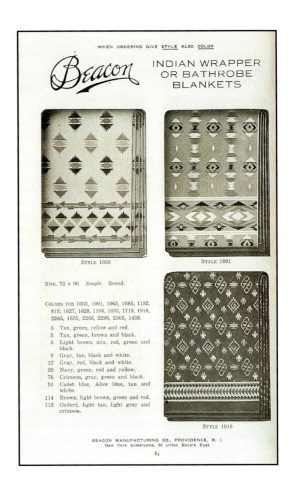

Beacon INDIAN WRAPPER OR BATHROBE BLANKETS

STYLE 1035

STYLE 1091

Size, 72 x 90. *Single.* Boxed.

COLORS FOR 1035, 1091, 1065, 1085, 1192,
819, 1627, 1628, 1108, 1603, 1718, 1916,
2265, 1655, 2268, 2295, 2303, 1438.

3   Tan, green, yellow and red.
5   Tan, green, brown and black.
6   Light brown mix, red, green and black.
9   Gray, tan, black and white.
12 Gray, red, black and white.
39 Navy, green, red and yellow.
76 Crimson, gray, green and black.
91 Cadet blue, Alice blue, tan and white.
114 Brown, light brown, green and red.
115 Oxford, light tan, light gray and crimson.

STYLE 1916

BEACON MANUFACTURING CO., PROVIDENCE, R. I.
New York Salesrooms, 50 Union Square East.
84

Beacon WRAPPER OR BATHROBE BLANKETS

STYLE 1729

STYLE 2299

STYLE 2270

Girdles, neck cords and frogs to match
wrapper blankets extra; per set, 25c.

Above are three-color designs.   *For descriptions and colors see page 77.*

BEACON MANUFACTURING CO., PROVIDENCE, R. I.
New York Salesrooms, 50 Union Square East.
81

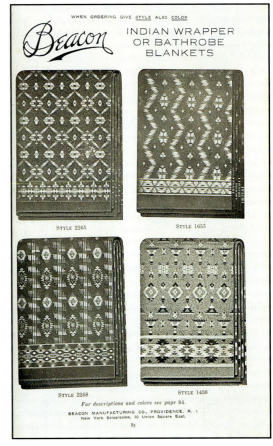

Beacon INDIAN WRAPPER OR BATHROBE BLANKETS

STYLE 2265

STYLE 1655

STYLE 2268

STYLE 1438

For descriptions and colors see page 84.

BEACON MANUFACTURING CO., PROVIDENCE, R. I.
New York Salesrooms, 50 Union Square East.
85

48   From the 1917-1918 Fall Winter Catalog

Beacon PLAID COUCH BLANKETS

STYLE S R 1

STYLE S R 4

STYLE S R 2

STYLE S R 3

Size 70 x 84. *Single.* Boxed.

COLORS FOR S R 1 AND S R 4:
No. 1 Dark green, black, light brown and white.
2 Dark green, light brown, gray and white.
3 Navy, light brown, green and white.
4 Maroon, light brown, green and white.
5 Brown, light brown, maroon and white.
6 Black, green, light brown and white.
7 Green, light brown, red and white.
White appears in small spots only.

COLORS FOR S R 2 AND S R 3:
No. 1 Dark green, black and white.
2 Dark green, light brown and white.
3 Navy blue, light brown and white.
4 Maroon, light brown and white.
5 Brown, light brown and white.
6 Black, dark green and white.

BEACON MANUFACTURING CO., PROVIDENCE, R. I.
New York Salesrooms, 50 Union Square East.
107

Beacon INDIAN BATHROBE BLANKETS

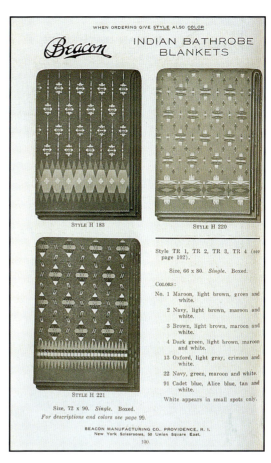

STYLE H 183

STYLE H 220

STYLE H 221

Size, 72 x 90. *Single.* Boxed.
For descriptions and colors see page 99.

Style TR 1, TR 2, TR 3, TR 4 (see page 102).

Size, 66 x 80. *Single.* Boxed.

COLORS:

No. 1 Maroon, light brown, green and white.

2 Navy, light brown, maroon and white.

3 Brown, light brown, maroon and white.

4 Dark green, light brown, maroon and white.

13 Oxford, light gray, crimson and white.

22 Navy, green, maroon and white.

91 Cadet blue, Alice blue, tan and white.

White appears in small spots only.

BEACON MANUFACTURING CO., PROVIDENCE, R. I.
New York Salesrooms, 50 Union Square East.
100.

Beacon INDIAN BATHROBE BLANKETS

STYLE H 102

STYLE H 84

STYLE H 104

STYLE H 145

Styles H 102, H 84, H 104, H 145, H 183, H 220, H 221.
Size, 72 x 90. *Single.* Boxed. COLORS:

No. 3 Tan, green, yellow and red.
5 Tan, green, brown and black.
6 Light brown, red, green and black.
9 Gray, tan, black and white.
12 Gray, red, black and white.

22 Navy, green, red and yellow.
76 Crimson, gray, green and black.
91 Cadet blue, Alice blue, tan and white.
114 Brown, light brown, green and red.
115 Oxford, light brown, light gray and crimson.

BEACON MANUFACTURING CO., PROVIDENCE, R. I.
New York Salesrooms, 50 Union Square East.
99

Beacon INDIAN WRAPPER OR BATHROBE BLANKETS

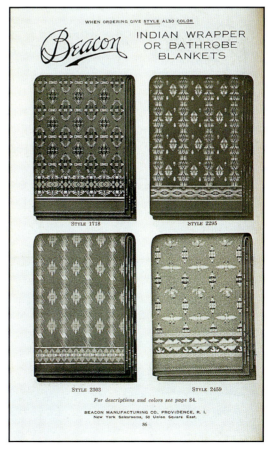

STYLE 1718

STYLE 2295

STYLE 2303

STYLE 2459

For descriptions and colors see page 84.

BEACON MANUFACTURING CO., PROVIDENCE, R. I.
New York Salesrooms, 50 Union Square East.
86

From the 1917-1918 Fall Winter Catalog

49

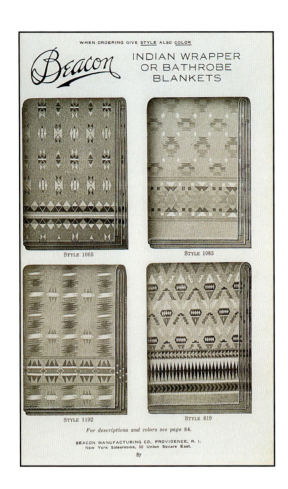

Beacon INDIAN WRAPPER OR BATHROBE BLANKETS

STYLE 1065   STYLE 1085

STYLE 1192   STYLE 819

For descriptions and colors see page 84.

BEACON MANUFACTURING CO., PROVIDENCE, R. I.
New York Salesrooms, 50 Union Square East.

87

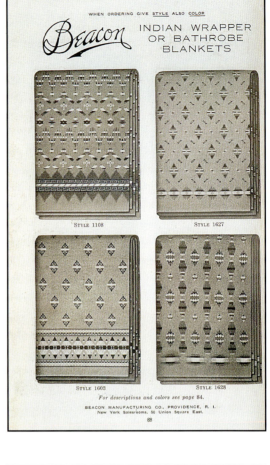

Beacon INDIAN WRAPPER OR BATHROBE BLANKETS

STYLE 1108   STYLE 1627

STYLE 1603   STYLE 1628

For descriptions and colors see page 84.

BEACON MANUFACTURING CO., PROVIDENCE, R. I.
New York Salesrooms, 50 Union Square East.

88

Beacon INDIAN WRAPPER OR BATHROBE BLANKETS

STYLE 1065   STYLE 1085

STYLE 1192   STYLE 819

For descriptions and colors see page 84.

BEACON MANUFACTURING CO., PROVIDENCE, R. I.
New York Salesrooms, 50 Union Square East.

87

Beacon CRIB BLANKETS

STYLE N 1   STYLE N 2

Styles N 1 and N 2 Crib Blankets. Pairs. Made in white with pink or blue
borders; also in blue or pink with white borders. Size, 36 x 50. Pairs boxed.
Bound with 1½-inch blanket binding.

INDIAN CRIB BLANKETS

STYLE IND. CRIB 1   STYLE IND. CRIB 4

Style Ind. Crib 1 and 4. Size, 36 x 50. Single. Boxed.
Colors.

3 tan, green, red and yellow.   39 navy, red, green and yellow.
6 light brown mix, red, green and black.   116 light blue, white, tan and Alice blue.
9 gray, tan, black and white.   117 pink, white, tan and light green.
12 gray, red, black and white.   118 Oxford, tan, light gray and purple.

BEACON MANUFACTURING CO., PROVIDENCE, R. I.
New York Salesrooms, 50 Union Square East.

56

From the 1917-1918 Fall Winter Catalog

Four colors were the maximum that could be woven into a blanket, at Beacon, but this blanket looks as if it has many more, "c. 1920s." Courtesy of Sam and Denise Kennedy of Cisco's, Coeur d'Alene, Idaho.

Strong vivid colors and dense napping make a great combination on this 1920 period Indian blanket. Courtesy of Charles D. Owen III.

An attractive color combination with a bow and arrow motif, "c. 1922." Author's collection. $150-225

Beautiful graphics and color arrangement give this excellent blanket a third dimensional affect, early-1920s. Courtesy of Laura Fisher/Antique Quilts and Americana.

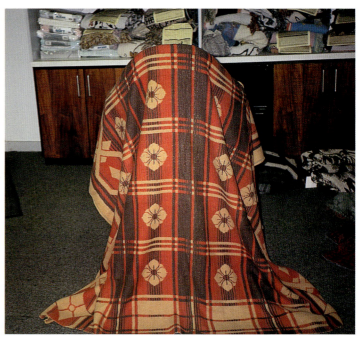

Tan, maroon and red combine pleasantly on this very typical Beacon blanket, "c. 1920s." Courtesy of Beacon Manufacturing Company.

This design appears to be an Ombre but closer examination reveals it's a forerunner, early 1920s. Courtesy of Charles D. Owen III.

A 1924 Yukon single blanket originally available in six different color combinations, Authors collection, $50-125.

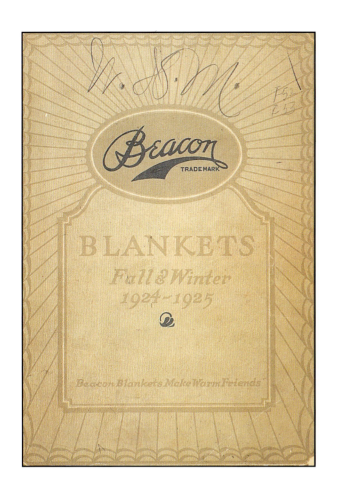

BATHROBE BLANKETS

Style 1301

Style 2244

Style 1374

Style 3837

Bathrobe or Wrapper Blankets. Size, 72 x 90. Single, boxed

Medium gray and light blue	5046	Navy and red	2324
Oxford gray and red	8024	Brown and tan	1825
Pearl and pink	4111	Rose and white	16w
Alice and white	22w	Dark green and red	3224
Copenhagen and tan	5525		

ROBE CLOTH

Pattern 2294

Pattern 1342

Pattern 2265

Pattern 855

Pattern 2292

Pattern 1627

Beacon Robe Flannel, 36 inches wide; 25-yard pieces

Light brown, red, green and black	6	Gray, red, black and tan	12
Tan, green, brown and black	5	Navy, green, red and yellow	39
Copenhagen, Alice, tan and white	91	Navy, green, red and tan	193
Brown, light brown, green and red	114		

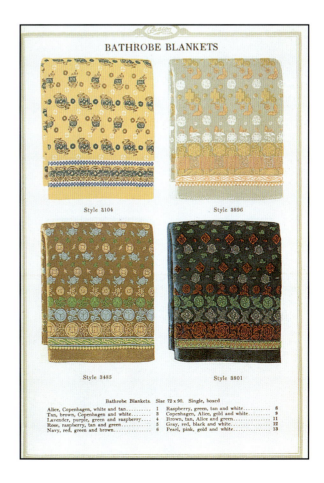

BATHROBE BLANKETS

Style 3104

Style 3896

Style 3485

Style 3801

Bathrobe Blankets. Size 72 x 90. Single, boxed

Alice, Copenhagen, white and tan	1	Raspberry, green, tan and white	8
Tan, brown, Copenhagen and white	3	Copenhagen, Alice, gold and white	9
Lavender, purple, green and raspberry	4	Brown, tan, Alice and green	11
Rose, raspberry, tan and green	5	Gray, red, black and white	12
Navy, red, green and brown	6	Pearl, pink, gold and white	13

BATHROBE BLANKETS

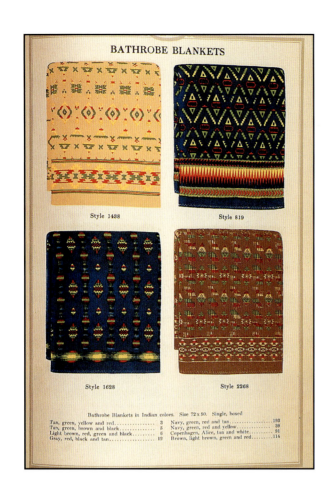

Style 1438

Style 819

Style 1628

Style 2268

Bathrobe Blankets in Indian colors. Size 72 x 90. Single, boxed

Tan, green, yellow and red	3	Navy, green, red and tan	193
Tan, green, brown and black	5	Navy, green, red and yellow	39
Light brown, red, green and black	6	Copenhagen, Alice, tan and white	91
Gray, red, black and tan	12	Brown, light brown, green and red	114

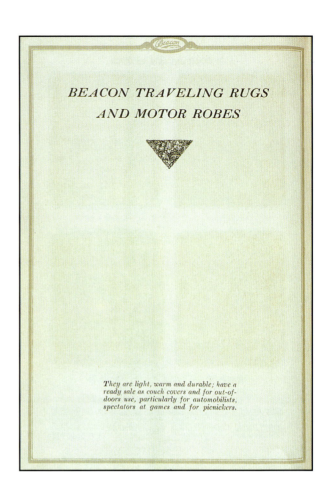

BEACON TRAVELING RUGS AND MOTOR ROBES

They are light, warm and durable; have a ready sale as couch covers and for out-of-doors use, particularly for automobilists, spectators at games and for picnickers.

BATHROBE BLANKETS

Style 1108

Style 1627

Style 2792

Style 1916

Bathrobe Blankets in Indian colors. Size 72 x 90. Single, boxed

Tan, green, yellow and red	3	Navy, green, red and tan	193
Tan, green, brown and black	5	Navy, green, red and yellow	39
Light brown, red, green and black	6	Copenhagen, Alice, tan and white	91
Gray, red, black and tan	12	Brown, light brown, green and red	114

MOTOR AND TRAVELING ROBES

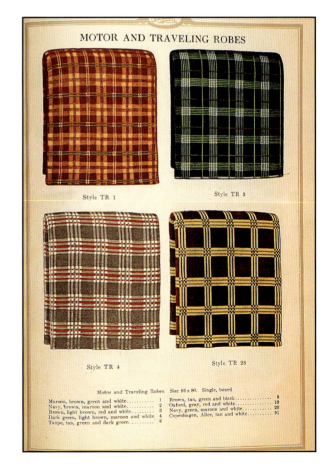

Style TR 1

Style TR 3

Style TR 4

Style TR 28

Motor and Traveling Robes. Size 66 x 80. Single, boxed

Maroon, brown, green and white	1	Brown, tan, green and black	8
Navy, brown, maroon and white	2	Oxford, gray, red and white	18
Brown, light brown, red and white	3	Navy, green, maroon and white	20
Dark green, light brown, maroon and white	4	Copenhagen, Alice, tan and white	91
Taupe, tan, green and dark green	6		

54

RAINBOW BLANKETS

Rainbow 4   Rainbow 2

Rainbow 3   Rainbow 5

Rainbow Blankets are attractive plaids of convenient size and weight for use as couch throws.
Size 60 x 80. Bound around with 2-inch tape

Dark green, green mix, light green	1	Navy, Copenhagen, Alice	3
Brown, drab, light tan	2	Maroon, drab, tan	4

INDIAN BLANKETS

Style Wigwam 20   Style Wigwam 21

Style Wigwam 22   Style Wigwam 23

Wigwam Blankets. Size 60 x 80. Single, boxed. Bound around with 2-inch tape

Taupe, tan, brown and black	141	Copenhagen, Alice, tan and brown	192
Tan, green, red and yellow	3	Navy, red, green and yellow	39
Tan, green, brown and black	5	Navy, red, green and tan	193
Light brown, red, green and black	6	Pearl, orange, green and black	191
Light brown, red, yellow and black	15	Light green, orange, Alice and purple	184
Gray, red, black and tan	12	Brown, maroon, dark green and green	157
Copenhagen, Alice, tan and white	91	Brown, tan, green and red	114

BEACON INDIAN BLANKETS

DISTINCTIVE AND DECORATIVE

Patterns adapted from the best in American Indian design. A great variety of Patterns in bright colors.

For display purposes and for window dressing Beacon Indian Blankets are unexcelled, and many pleasing eye-catching displays can be arranged in connection with our unique show cards.

INDIAN BLANKETS

Style Wigwam 3   Style Wigwam 16

Style Wigwam 17   Style Wigwam 19

Wigwam Blankets. Size 60 x 80. Single, boxed. Bound around with 2-inch tape

Taupe, tan, brown and black	141	Copenhagen, Alice, tan and brown	192
Tan, green, red and yellow	3	Navy, red, green and yellow	39
Tan, green, brown and black	5	Navy, red, green and tan	193
Light brown, red, green and black	6	Pearl, orange, green and black	191
Light brown, red, yellow and black	15	Light green, orange, Alice and purple	184
Gray, red, black and tan	12	Brown, maroon, dark green and green	157
Copenhagen, Alice, tan and white	91	Brown, tan, green and red	114

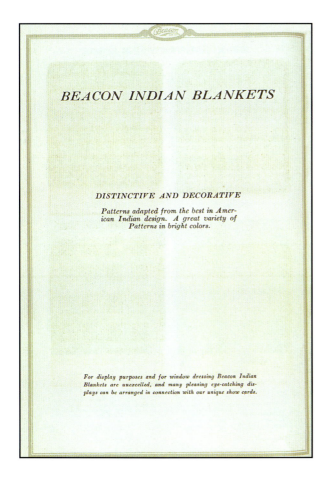

Style Sachem 5   Style Sachem 7

Style Sachem 8   Style Sachem 9

Sachem Blankets. Size 66 x 80. Single, boxed. Bound around with 2-inch tape

Taupe, tan, brown and black	141	Copenhagen, Alice, tan and brown	192
Tan, green, red and yellow	3	Navy, red, green and yellow	39
Tan, green, brown and black	5	Navy, red, green and tan	193
Light brown, red, green and black	6	Pearl, orange, green and black	191
Light brown, red, yellow and black	15	Light green, orange, Alice and purple	184
Gray, red, black and tan	12	Brown, maroon, dark green and green	157
Copenhagen, Alice, tan and white	91	Brown, tan, green and red	114

Style Sachem 19   Style Sachem 20

Style Sachem 26   Style Sachem 27

Sachem Blankets. Size 66 x 80. Single, boxed. Bound around with 2-inch tape

Taupe, tan, brown and black	141	Copenhagen, Alice, tan and brown	192
Tan, green, red and yellow	3	Navy, red, green and yellow	39
Tan, green, brown and black	5	Navy, red, green and tan	193
Light brown, red, green and black	6	Pearl, orange, green and black	191
Light brown, red, yellow and black	15	Light green, orange, Alice and purple	184
Gray, red, black and tan	12	Brown, maroon, dark green and green	157
Copenhagen, Alice, tan and white	91	Brown, tan, green and red	114

Style Sachem 1   Style Sachem 2

Style Sachem 3   Style Sachem 4

Sachem Blankets. Size 66 x 80. Single, boxed. Bound around with 2-inch tape

Taupe, tan, brown and black	141	Copenhagen, Alice, tan and brown	192
Tan, green, red and yellow	3	Navy, red, green and yellow	39
Tan, green, brown and black	5	Navy, red, green and tan	193
Light brown, red, green and black	6	Pearl, orange, green and black	191
Light brown, red, yellow and black	15	Light green, orange, Alice and purple	184
Gray, red, black and tan	12	Brown, maroon, dark green and green	157
Copenhagen, Alice, tan and white	91	Brown, tan, green and red	114

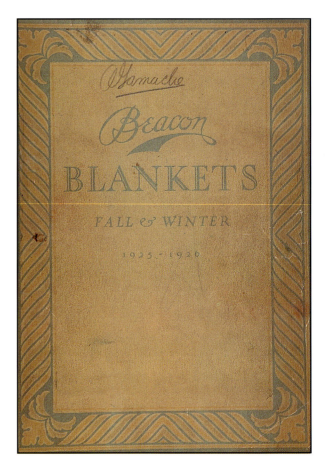

Damache

Beacon

BLANKETS

FALL & WINTER

1925-1926

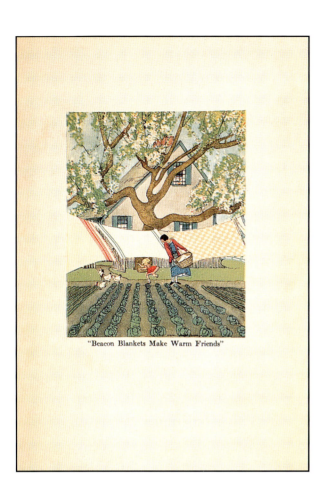

"Beacon Blankets Make Warm Friends"

INDIAN BLANKETS

Style Sachem 19

Style Wigwam 28

Style Sachem 34

Style Sachem 27

Sachem Blankets. Size 66 x 80. Bound around with 2-inch tape. Single, boxed.

Tan, green, red and yellow	3	Copenhagen, Alice, tan and white	91
Tan, green, brown and black	5	Copenhagen, Alice, tan and brown	192
Taupe, tan, brown and black	141	Navy, red, green and yellow	89
Taupe, orange, green and black	205	Navy, red, green and tan	193
Light brown, red, green and black	6	Pearl, orange, green and black	191
Light brown, red, yellow and black	15	Brown, maroon, dark green and green	157
Gray, red, black and tan	12	Brown, tan, green and red	114

40

INDIAN BLANKETS

Style Wigwam 24

Style Wigwam 28

Style Wigwam 32

Style Wigwam 33

Wigwam Blankets. Size 60 x 80. Single, boxed. Bound around with 2-inch tape.

Tan, green, red and yellow	3	Copenhagen, Alice, tan and white	91
Tan, green, brown and black	5	Copenhagen, Alice, tan and brown	192
Taupe, tan, brown and black	141	Navy, red, green and yellow	89
Taupe, orange, green and black	205	Navy, red, green and tan	193
Light brown, red, green and black	6	Pearl, orange, green and black	191
Light brown, red, yellow and black	15	Brown, tan, green and red	114
Gray, red, black and tan	12	Brown, maroon, dark green and green	157

38

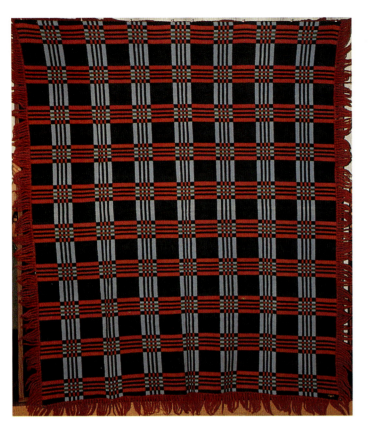

The early shawl with Beacon's classic orange fringe (and there was lots of it), "c. 1919." Courtesy of Sam and Denise Kennedy of Cisco's, Coeur d'Alene, Idaho.

Vibrant colors of a black, red and blue shawl. Courtesy of Laura Fisher/Antique Quilts and Americana.

Artistic small scale Indian design on a fringed blanket shawl, "c. 1918."
Courtesy of Sam and Denise Kennedy of Cisco's, Coeur d'Alene, Idaho.

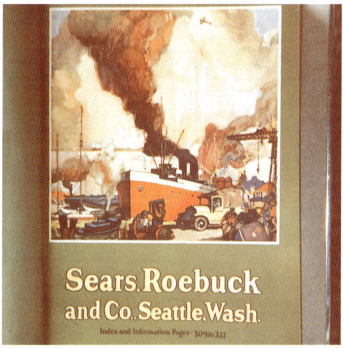

Fringed shawl with pleasant Indian motif and orange fringe early-1920s. Courtesy of Bearwallow Mountain Traders, Judy Hudson & Norwood Barnes.

Indian design fringed shawl with classic orange fringe. Excellent design and use of color, "c. 1924."
Courtesy of Bearwallow Mountain Traders, Judy Hudson & Norwood Barnes.

Two For The Price of One, Double Cloth

There are two sides to every blanket. The different appearances of the two sides are created by two sets of filling threads and one set of warp threads; this is a "double cloth." The warp is the yarn that runs lengthwise and the filling is yarn that runs crosswise. Fillings are carried over and under the warp yarns creating this effect. The positive side of a blanket is where the pattern appears the boldest, called the "patternization" side. The reverse side appears negative and the color placement is opposite or reversed, known as the "colorization" side. It is like getting two blankets for the price of one.

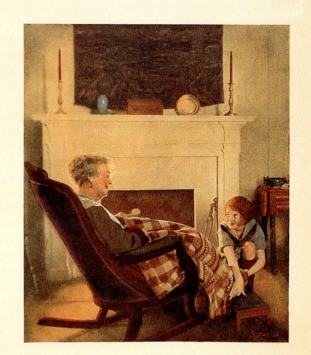

"Beacon Blankets Make Warm Friends"

The patternization side of an early Indian design blanket, "c. 1915." Courtesy of Gary Holt/Steve Christianson, Gary Holt collection.

The colorization side of the same blanket, "c. 1915." Courtesy of Gary Holt/Steve Christianson, Gary Holt collection.

Patternization side of a pleasing combination of two Indian design patterns, late-1930s. Courtesy of Bearwallow Mountain Traders, Judy Hudson & Norwood Barnes.

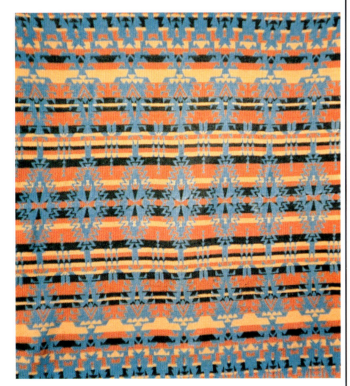

Colorization side of this unique blanket. The opposite colors of the color spectrum give the design depth, late-1930.

Ombre Period: 1926-1950

The "Ombre" Effect

Charles Owen Dexter invented a complex process to enhance the attractiveness of Beacon blankets and robe cloth. This design process known as "*Ombre*" gave the designs a shading quality or third-dimensional effect, or, as stated in the company's 1928 catalog: "A silky rainbow effect obtained by shading of rich colors." This process was incorporated in the design of many of Beacon's fabrics, especially ones with Indian designs, and gave Beacon products a look that went beyond the four color limit of the Jacquard loom.

The *Ombre* process encompassed the utilization of predetermining patterns and color layouts thereby enabling a complete blending of color in a continuous spectrum. Previously this had been achievable only by hand spinning and weaving. This unique process would be very costly to duplicate today, as it is extremely labor intensive. The *Ombre* process is one of the more obvious and desirable characteristics that made Beacon blankets the industry leader.

The Ombre Process

Cotton fiber was known as "raw stock" or "kettle dyed". The fiber was dyed under pressure in two thousand pound kettles enabling the dye to penetrate all areas of the "fiber cake" or "kettle load".

The carding/spinning system required the removal of the conventional card hopper and pin aprons. Normally, there was a feed hopper with two aprons, one with slats to feed the fiber and one with pins to deliver the fiber into the carding system. This was designed for a single solid color fiber.

The hopper was replaced with a long feed apron. The colors were placed on the apron in order of color requirements. In processing an *Ombre* design the colors were arranged in light, medium and dark shades: light red melded to medium red and then to dark red, before reversing to medium red and back to light red.

This method was both simple and complicated. The colors were weighed and placed on a long apron enabling the fiber to be fed into the card to produce an exact length of roving to create a bobbin or quill totaling 180 yards of yarn. The fact that various colors were being fed side by side allowed a mechanical mixing of color that resulted in a harmonious mixture of shades between the various colors, thus resulting in a true *Ombre* or color blending.

The weaving system used bobbins that were preselected to assure the correct color rotation in the woven blanket. Although the Crompton and Knowles looms had only four filling positions or boxes, the loom was set to automatically change or stop off to allow a color change or placement. This enabled an unlimited variety of color combinations, various border patterns and/or colors, and symmetry of design and pattern. Since these blankets were processed in various widths, bobbin selections had to be predetermined for each size, pattern and coloration.

1926 Bath Robe Blankets. Center item shows the first appearance of a Beacon Ombre fabric. $5.95 each. A bath robe blanket could be used as a blanket, a wrap or be made into a bathrobe. Left and right models, $5.15 each. Neck and waist cords were included. Courtesy of Sears, Roebuck & Co. archives, reprinted by special arrangement and protected by copyright. No duplication is permitted.

Napping or fiber raising was the next step in making the fabric into blankets. This was a continuous fabric process where the loom-woven rolls were run through a napping machine consisting of a series of wire rollers with alternating counter and clockwise pile rolls. The rolls were adjusted individually to gently raise the cotton fibers resulting in a very soft blanket pile covering both sides of the blanket. This was an important process, as it enhanced the color blending and brightness of the blanket as well as the soft finish that Beacon had become famous for.

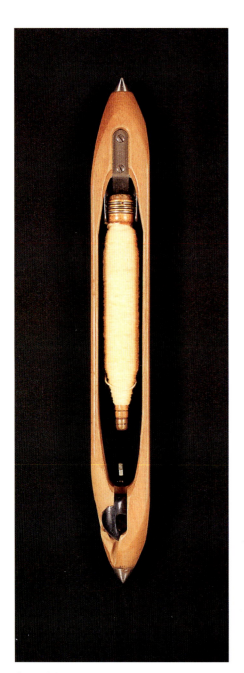

One of the original shuttles used in the *Ombre* process. The bobbin in the center held 180 yards of yarn. Courtesy of Charles D. Owen Jr. IV.

Smoke Signals. Beacon's 1927 advertising book explained, "Taken from an oil painting recently made for us showing an Indian signaling from a high point with smoke signals made by covering a smoldering fire with his blanket. The blanket is our style Wigwam No. 75." Courtesy of Beacon Manufacturing Co.

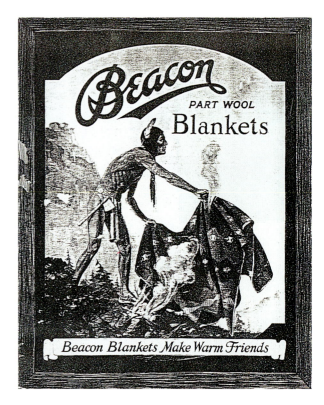

After the napping and cleaning, the blanket fabric went to the hemming and binding department where the long continuous material was cut into blankets. Depending on the style of the finished blankets, some would get bindings made of silk or mercerized cotton or satin, and some were whip stitched on the ends. Some of the designs included a yarn-fringed border, creating a shawl effect.

Beacon had succeeded in producing a low-cost, attractive cotton blanket and in some products cotton and wool blends were manufactured containing between five and eighty five percent wool.

Patternization side of an exciting black fringe shawl, late-1920s.

An outstanding example of the *Ombre* effect on a 1930s shawl/blanket. Courtesy of David W. Schutzler, Enumclaw, Washington.

Colorization side....it's like getting two shawls for the price of one. Courtesy of Bearwallow Mountain Traders, Judy Hudson & Norwood Barnes.

Wardrobe closets brim full of beautiful Beacon blankets. Courtesy of Sam and Denise Kennedy of Cisco's, Coeur d'Alene, Idaho.

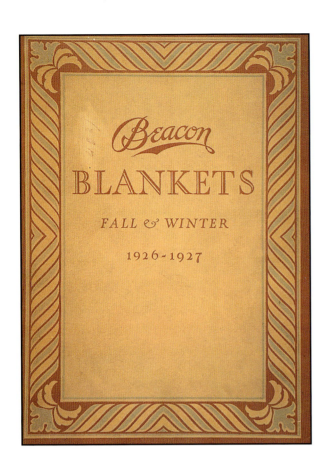

Beacon

BLANKETS

FALL & WINTER

1926-1927

Distinctive and decorative. Patterns adapted from the best in American Indian design. A great variety of patterns in bright colors.

For display purposes and for window dressing Beacon Indian Blankets are unexcelled, and many pleasing, eye-catching displays can be arranged in connection with our unique showcards.

BATHROBE BLANKETS

In the following eleven pages are illustrated bathrobe blankets in patterns which we endeavor to carry in stock in complete ranges of colors. The size is two yards wide by two and a half yards long— 72 x 90—ample cloth from which to make one bathrobe.

Blankets in two weights are illustrated—the H quality being heavier. All are single and boxed separately, with or without girdle set.

Style Wigwam 16   Style Wigwam 23

Style Wigwam 3   Style Wigwam 22

WIGWAM BLANKETS. *Size 60 x 80. Bound all around with 1½-inch tape. Single, boxed*

Tan, green, red and yellow	3	Copenhagen, Alice, tan and white	91
Tan, green, brown and black	5	Copenhagen, Alice, tan and brown	192
Taupe, tan, brown and black	141	Navy, red, green and yellow	39
Taupe, orange, green and black	205	Navy, red, green and tan	193
Light brown, red, green and black	6	Pearl, orange, green and black	191
Light brown, red, yellow and black	15	Brown, maroon, dark green and green	137
Gray, red, black and tan	12	Brown, tan, green and red	114

64

Style Sachem 20   Style Sachem 5

Style Sachem 3   Style Sachem 2

SACHEM BLANKETS. *Size 66 x 80. Bound all around with 1½-inch tape. Single, boxed*

Tan, green, red and yellow	3	Copenhagen, Alice, tan and white	91
Tan, green, brown and black	5	Copenhagen, Alice, tan and brown	192
Taupe, tan, brown and black	141	Navy, red, green and yellow	39
Taupe, orange, green and black	205	Navy, red, green and tan	193
Light brown, red, green and black	6	Pearl, orange, green and black	191
Light brown, red, yellow and black	15	Brown, maroon, dark green and green	157
Gray, red, black and tan	12	Brown, tan, green and red	114

Style Sachem 33   Style Sachem 34

Style Sachem 19   Style Sachem 27

SACHEM BLANKETS. *Size 66 x 80. Bound all around with 1½-inch tape. Single, boxed*

Tan, green, red and yellow	3	Copenhagen, Alice, tan and white	91
Tan, green, brown and black	5	Copenhagen, Alice, tan and brown	192
Taupe, tan, brown and black	141	Navy, red, green and yellow	39
Taupe, orange, green and black	205	Navy, red, green and tan	193
Light brown, red, green and black	6	Pearl, orange, green and black	191
Light brown, red, yellow and black	15	Brown, tan, green and red	114
Gray, red, black and tan	12	Brown, maroon, dark green and green	157

Style Wigwam 33   Style Wigwam 24

Style Wigwam 27   Style Wigwam 17

WIGWAM BLANKETS. *Size 60 x 80. Bound all around with 1½-inch tape. Single, boxed*

Tan, green, red and yellow	3	Copenhagen, Alice, tan and white	91
Tan, green, brown and black	5	Copenhagen, Alice, tan and brown	192
Taupe, tan, brown and black	141	Navy, red, green and yellow	39
Taupe, orange, green and black	205	Navy, red, green and tan	193
Light brown, red, green and black	6	Pearl, orange, green and black	191
Light brown, red, yellow and black	15	Brown, maroon, dark green and green	157
Gray, red, black and tan	12	Brown, tan, green and red	114

Style Wigwam 32   Style Wigwam 20

Style Wigwam 19   Style Wigwam 28

WIGWAM BLANKETS. *Size 60 x 80. Bound all around with 1½-inch tape. Single, boxed*

Tan, green, red and yellow	3	Copenhagen, Alice, tan and white	91
Tan, green, brown and black	5	Copenhagen, Alice, tan and brown	192
Taupe, tan, brown and black	141	Navy, red, green and yellow	39
Taupe, orange, green and black	205	Navy, red, green and tan	193
Light brown, red, green and black	6	Pearl, orange, green and black	191
Light brown, red, yellow and black	15	Brown, maroon, dark green and green	157
Gray, red, black and tan	12	Brown, tan, green and red	114

TOPAZ AND YUKON
BLANKETS

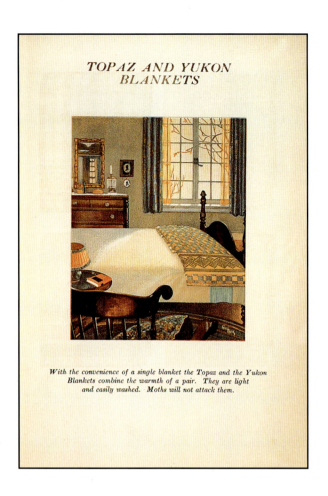

*With the convenience of a single blanket the Topaz and the Yukon
Blankets combine the warmth of a pair. They are light
and easily washed. Moths will not attack them.*

PLAIN AND PLAID BLANKETS

*At a slight additional cost per pair we will cut double blankets in
two and bind the ends. If desired in this way please order,
"cut single and bound."*

BEACON TRAVELING RUGS
AND MOTOR ROBES

*They are light, warm and durable; have a ready sale as couch covers
and for out-of-doors use, particularly for automobilists,
spectators at games and for picnickers.*

JACQUARD COMFORTABLES

*In all particulars unmistakably superior as extra covers and spread
blankets. Extra thick and warm, yet light in weight and easily
handled, they are durable and will outwear any stuffed comforter or
quilt, are mothproof and can be washed. We are justly proud of
these comfortables. They please the most discriminating tastes.*

*Our customers can secure estimates from us for furnishing special
blankets for institutions, hotels and colleges. Crests, coats-of-arms or
names can be woven into blankets if desired.*

Excellent Examples of Ombre Blankets

Courtesy of Sam and Denise Kennedy of Cisco's, Coeur d'Alene, Idaho.

Courtesy of Charles D. Owen III.

Courtesy of Laura Fisher/Antique Quilts and Americana.

Courtesy of Sam and Denise Kennedy of Cisco's, Coeur d'Alene, Idaho.

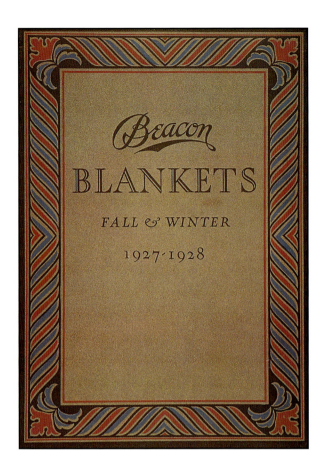

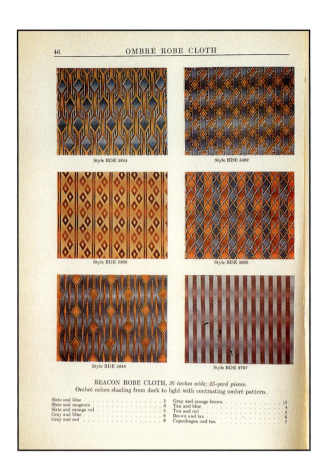

Style BDE 5244   Style BDE 5592

Style BDE 5268   Style BDE 5506

Style BDE 5249   Style BDE 2787

BEACON ROBE CLOTH, *36 inches wide; 25-yard pieces.*
Ombré colors shading from dark to light with contrasting ombré pattern.

Slate and blue	1	Gray and orange brown	10
Slate and magenta	2	Tan and blue	4
Slate and orange red	3	Tan and red	5
Gray and blue	8	Brown and tan	6
Gray and red	9	Copenhagen and tan	7

Style E 2294 (27 inches)
Style EB 2294 (36 inches)

Style E 2292 (27 inches)
Style EB 2292 (36 inches)

Style E 1027 (27 inches)
Style EB 1027 (36 inches)

Style E 855 (27 inches)
Style EB 855 (36 inches)

Style E 2265 (27 inches)
Style EB 2265 (36 inches)

Style E 4093 (27 inches)
Style EB 4093 (36 inches)

BEACON ROBE CLOTH, *27 inches wide, 30-yard pieces; 36 inches wide, 25-yard pieces.*

Tan, green, brown and black	5	Gray, red, black and tan	12
Taupe, tan, brown and black	141	Copenhagen, Alice, tan and brown	192
Light brown, red, green and black	6	Navy, green, red and yellow	39
Brown, light brown, green and red	114	Navy, green, red and tan	193
Pearl, orange, green and black	210		

Style BDE 5921   Style BDE 5938

Style BDE 5976   Style BDE 6034

Style BDE 6195   Style BDE 6242

BEACON ROBE CLOTH, *36 inches wide; 25-yard pieces.*
Ombré colors shading from dark to light with contrasting ombre pattern.

Slate and blue	1	Gray and orange brown	10
Slate and magenta	2	Tan and blue	4
Slate and orange red	3	Tan and red	5
Gray and blue	8	Brown and tan	6
Gray and red	9	Copenhagen and tan	7

Style Yukon 29   Style Yukon 9

Style Yukon 12   Style Yukon 5

YUKON BLANKETS. Combining the convenience of a single blanket with the warmth of a pair.
Sizes 66 x 84 and 72 x 84. Ends bound with 3-inch sateen. Single.

Alice, tan and buff	22T	Gold, tan and buff	19
Alice, light blue and buff	22B	Light brown, taupe and tan	62
Rose, pink and buff	16P	Brown, taupe and tan	18
Rose, Alice and buff	16B	Slate, taupe and tan	98
Lavender, tan and buff	13	Copenhagen, Alice and buff	55B
Pearl, pink and buff	80P	Copenhagen, taupe and tan	55T
Pearl, blue and buff	80B	Maroon, red and pearl	78

Style Agawam 23   Style Agawam 19

Style Agawam 20   Style Agawam 25

AGAWAM BLANKETS. *Size 60 x 80. Bound all around with 1½-inch tape. Single, boxed.*

Tan, green, brown and black	5	Copenhagen, Alice, tan and white	91
Taupe, tan, brown and black	141	Copenhagen, Alice, tan and brown	192
Taupe, orange, green and black	205	Navy, red, green and yellow	39
Light brown, red, green and black	6	Navy, red, green and tan	193
Brown, tan, green and red	114	Pearl, orange, green and black	191
Brown, maroon, dark green and green	157	Gray, red, black and tan	12

Style Huron 4   Style Huron 3

Style Huron 1   Style Huron 2

HURON BLANKETS. *Size 66 x 80. Bound all around with 1½-inch tape. Single, boxed.*
An unusual treatment of Indian designs in ombré colors.

Tan, blue, black and red	1	Blue, tan, orange and black	2
Tan, gray, yellow and blue	3	Blue, red, yellow and green	10
Tan, green, blue and orange	6	Slate, tan, green and black	8
Tan, red, black and green	9	Slate, magenta, orange and green	11
Brown, tan, yellow and blue	4	Gray, red, orange and green	7
Red, yellow, blue and black	5		

Style Wigwam 73   Style Wigwam 74

Style Wigwam 66   Style Wigwam 67

WIGWAM BLANKETS. *Size 66 x 80. Bound all around with 1½-inch tape. Single, boxed.*

Tan, green, brown and black	5	Copenhagen, Alice, tan and white	91
Taupe, tan, brown and black	141	Copenhagen, Alice, tan and brown	192
Taupe, orange, green and black	205	Navy, red, green and yellow	39
Light brown, red, green and black	6	Navy, red, green and tan	193
Brown, tan, green and red	114	Pearl, orange, green and black	191
Brown, maroon, dark green and green	157	Gray, red, black and tan	12

Style Wigwam 41   Style Wigwam 45

Style Wigwam 75   Style Wigwam 43

WIGWAM BLANKETS. *Size 66 x 80. Bound all around with 1½-inch tape. Single, boxed.*

Tan, green, brown and black	5	Copenhagen, Alice, tan and white	91
Taupe, tan, brown and black	141	Copenhagen, Alice, tan and brown	192
Taupe, orange, green and black	205	Navy, red, green and yellow	39
Light brown, red, green and black	6	Navy, red, green and tan	193
Brown, tan, green and red	114	Pearl, orange, green and black	191
Brown, maroon, dark green and green	157	Gray, red, black and tan	12

Style Agawam 22   Style Agawam 7

Style Agawam 21   Style Agawam 18

AGAWAM BLANKETS. *Size 60 x 80. Bound all around with 1½-inch tape. Single, boxed.*

Tan, green, brown and black	5	Copenhagen, Alice, tan and white	91
Taupe, tan, brown and black	141	Copenhagen, Alice, tan and brown	192
Taupe, orange, green and black	205	Navy, red, green and yellow	39
Light brown, red, green and black	6	Navy, red, green and tan	193
Brown, tan, green and red	114	Pearl, orange, green and black	191
Brown, maroon, dark green and green	157	Gray, red, black and tan	12

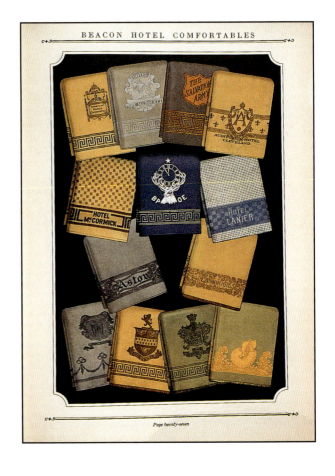

70

Style D 213 *Style D 1*

PLAIN COMFORTABLES. Single. With contrasting Jacquard borders. Sizes **60 x 82, 66 x 82 and 70 x 82.** Ends bound with 3-inch sateen. Colors:

Blue 22	Orchid . . . 71	Gold 19	Pearl 89
Rose 16	Green 86	Tan . . . 25	Tan with Brown 25T

Special Comfortables

NAMES, Crests or Coats-of-arms can be woven into the blankets if desired. Procure estimates from us for furnishing special blankets for institutions, hotels, steamships and colleges. Can be had with ends bound or hemmed.

Style Huron 4 *Style Huron 7*

Style Huron 1 *Style Huron 2*

HURON BLANKETS. Single. Size 66 x 80. Bound all around with 1½-inch tape. Boxed. An unusual treatment of Indian designs in ombré colors.

Tan, Blue, Black and Red . . . 1	Gray, Red, Orange and Green . . 7	
Blue, Tan, Orange and Black . . . 2	Pearl, Tan, Green and Black . . . 8	
Tan, Gray, Yellow and Blue . . . 3	Tan, Red, Black and Green . . . 9	
Brown, Tan, Yellow and Blue . . . 4	Blue, Red, Yellow and Green . . 10	
Red, Yellow, Blue and Black . . . 5	Pearl, Magenta, Orange and Green 11	
Tan, Green, Blue and Orange . . 6		

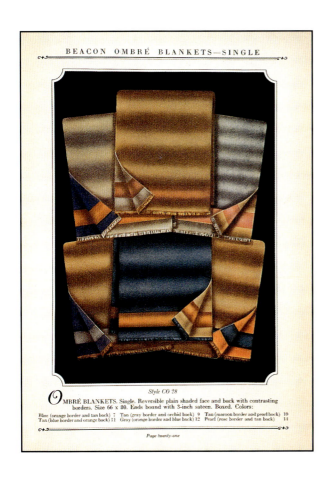

Style CO 28

OMBRÉ BLANKETS. Single. Reversible plain shaded face and back with contrasting borders. Size 66 x 80. Ends bound with 3-inch sateen. Boxed. Colors:

Blue (orange border and tan back) 7	Tan (gray border and orchid back) 9	Tan (maroon border and pearl back) 10
Tan (blue border and orange back) 11	Gray (orange border and blue back) 12	Pearl (rose border and tan back) 14

Style IX 3 *Style IX 4*

Style IX 20 *Style IX 6*

Style IX 5 *Style IX 23*

SINGLE CRIB BLANKETS. Line IX. Size 36 x 50. Boxed singly. Colors: Pink and White, Blue and White. Stitched or bound all around with 2-inch sateen.

Courtesy of Sam and Denise Kennedy of Cisco's, Coeur d'Alene, Idaho.

Courtesy of Beacon Manufacturing Company.

Courtesy of Laura Fisher/Antique Quilts and Americana.

Simulated Ombre. Courtesy of Sam and Denise Kennedy of Cisco's, Coeur d'Alene, Idaho.

Courtesy of Sam and Denise Kennedy of Cisco's, Coeur d'Alene, Idaho.

Courtesy of Laura Fisher/Antique Quilts and Americana.

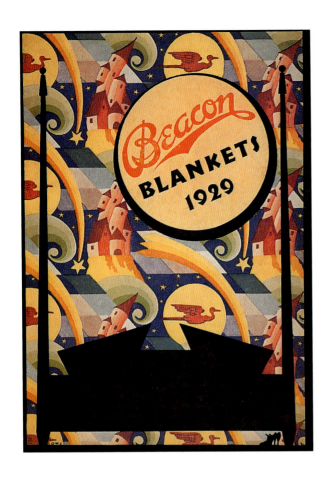

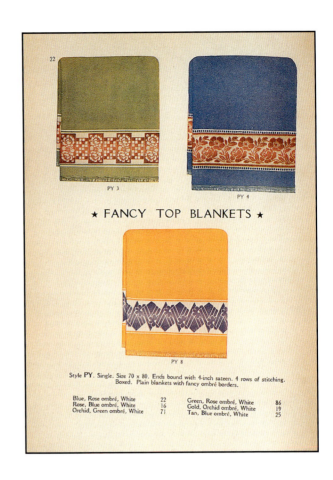

PY 3　　　　　PY 4

★ FANCY TOP BLANKETS ★

PY 8

Style **PY**. Single. Size 70 x 80. Ends bound with 4-inch sateen. 4 rows of stitching. Boxed. Plain blankets with fancy ombré borders.

Blue, Rose ombré, White	22	Green, Rose ombré, White	86
Rose, Blue ombré, White	16	Gold, Orchid ombré, White	19
Orchid, Green ombré, White	71	Tan, Blue ombré, White	25

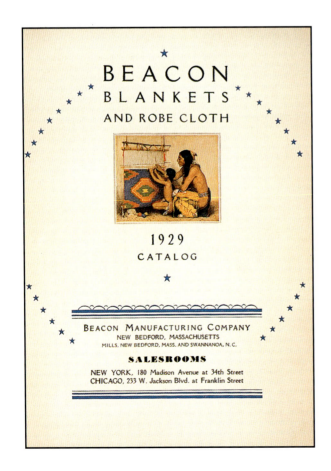

BEACON
BLANKETS
AND ROBE CLOTH

1929
CATALOG

★

BEACON MANUFACTURING COMPANY
NEW BEDFORD, MASSACHUSETTS
MILLS, NEW BEDFORD, MASS. AND SWANNANOA, N. C.

SALESROOMS

NEW YORK, 180 Madison Avenue at 34th Street
CHICAGO, 233 W. Jackson Blvd. at Franklin Street

YUKON 9　　　　　YUKON 14

★ YUKON ★ BLANKETS ★

YUKON 9

Style **YUKON**. Single. Size 70 x 80. Ends bound with 4-inch sateen. 4 rows of stitching. Combining the convenience of a single blanket with the warmth of a pair.

Blue, Tan, Buff	22			Copenhagen, Buff, Blue	55
Rose and Buff	91	Pearl, Pink, Buff	89	Maroon, Scarlet, Gray	78
Orchid, Tan, Buff	71	Light Brown, Taupe, Tan	62	Navy, Taupe, Buff	23
Gold, Tan, Buff	19	Brown, Taupe, Tan	18	Black, Red, Gray	35

Yukon 9 only—made in these additional colors:

Blue and White	22W			Green and White	86W
Rose and White	91W	Orchid and White	71W	Gold and White	19W

SIGNET 3.   SIGNET 7

★ SIGNET ★ BLANKETS ★

SIGNET 14   SIGNET 17

Style SIGNET. Single. Size 66 x 80. Ends bound with 3-inch sateen.

Blue, Tan, Buff	14	Green and White	86W
Blue and White	14W	Gold and White	19W
Rose, Tan, Buff	91	Brown, Taupe, Tan	18
Rose and White	91W	Copenhagen, Light Blue, Buff	55
Orchid, Tan, Buff	70	Maroon, Light Brown, Buff	78
Orchid and White	70W	Navy, Light Brown, Buff	23

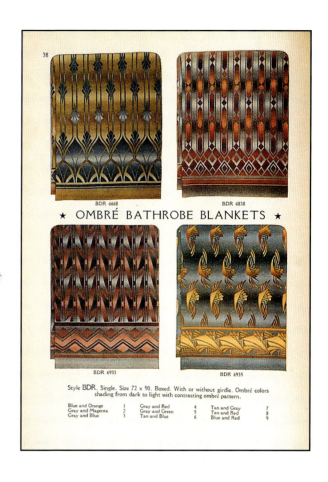

BDR 6668   BDR 6838

★ OMBRÉ BATHROBE BLANKETS ★

BDR 6933   BDR 6935

Style BDR. Single. Size 72 x 90. Boxed. With or without girdle. Ombré colors shading from dark to light with contrasting ombré pattern.

Blue and Orange	1	Gray and Red	4	Tan and Gray	7
Gray and Magenta	2	Gray and Green	5	Tan and Red	8
Gray and Blue	3	Tan and Blue	6	Blue and Red	9

TOPAZ 28   TOPAZ 31

★ TOPAZ ★ BLANKETS ★

TOPAZ 33

Style TOPAZ. Single. Size 70 x 80. Ends bound with 3-inch sateen. Attractive Plaids of convenient size and weight for use as extra bed coverings.

Topaz 28, 31 and 33

Blue, Tan, Buff	22			Copenhagen, Buff, Blue	55
Rose and Buff	91	Pearl, Pink, Buff	89	Maroon, Scarlet, Gray	78
Orchid, Tan, Buff	71	Lt. Brown, Taupe, Tan	62	Navy, Taupe, Buff	23
Gold, Tan, Buff	19	Brown, Taupe, Tan	18	Black, Red, Gray	35

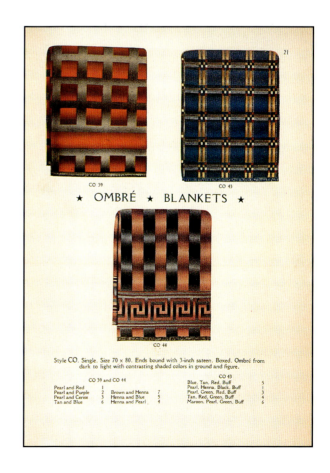

CO 39   CO 43

★ OMBRÉ ★ BLANKETS ★

CO 44

Style CO. Single. Size 70 x 80. Ends bound with 3-inch sateen. Boxed. Ombré from dark to light with contrasting shaded colors in ground and figure.

CO 39 and CO 44			CO 43	
Pearl and Red	1		Blue, Tan, Red, Buff	5
Pearl and Purple	2	Brown and Henna 7	Pearl, Henna, Black, Buff	1
Pearl and Cerise	3	Henna and Blue 5	Pearl, Green, Red, Buff	3
Tan and Blue	6	Henna and Pearl 4	Tan, Red, Green, Buff	4
			Maroon, Pearl, Green, Buff	6

STYLE H

★ PLAID ★ BLANKETS ★

Style H. Pairs. Size 70 x 80. Ends bound with 4-inch sateen. 4 rows of
stitching. 5-inch Block. Blue, Rose, Orchid, Green, Gold, Gray, Tan.

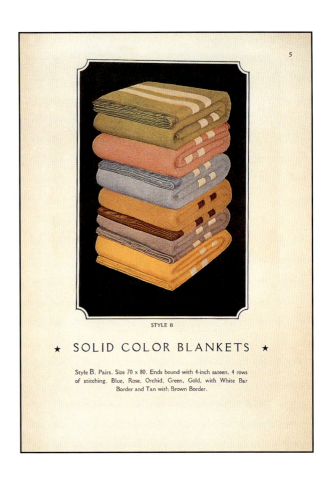

STYLE B

★ SOLID COLOR BLANKETS ★

Style B. Pairs. Size 70 x 80. Ends bound with 4-inch sateen. 4 rows
of stitching. Blue, Rose, Orchid, Green, Gold, with White Bar
Border and Tan with Brown Border.

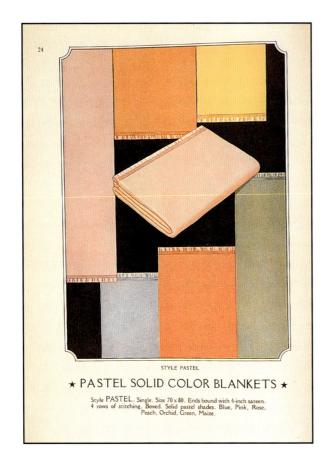

24

STYLE PASTEL

★ PASTEL SOLID COLOR BLANKETS ★

Style PASTEL. Single. Size 70 x 80. Ends bound with 4-inch sateen.
4 rows of stitching. Boxed. Solid pastel shades. Blue, Pink, Rose,
Peach, Orchid, Green, Maize.

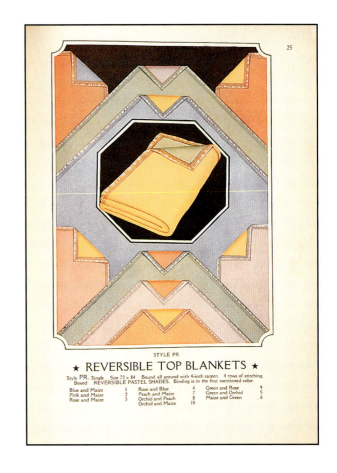

STYLE PR

★ REVERSIBLE TOP BLANKETS ★

Style PR. Single.   Size 72 x 84.   Bound all around with 4-inch sateen.   4 rows of stitching.
Boxed.   REVERSIBLE PASTEL SHADES.   Binding is in the first mentioned color.

Blue and Maize	1	Rose and Blue	4	Green and Rose	9
Pink and Maize	2	Peach and Maize	7	Green and Orchid	5
Rose and Maize	3	Orchid and Peach	8	Maize and Green	6
		Orchid and Maize	10		

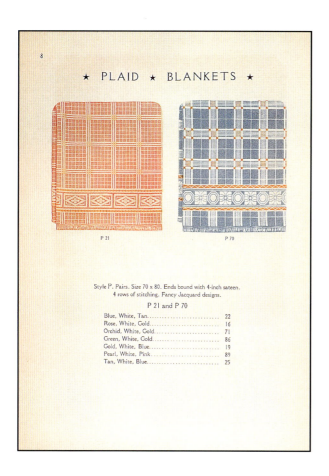

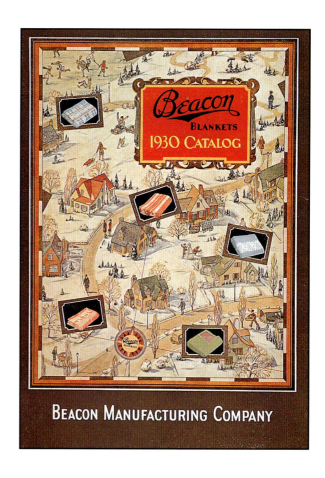

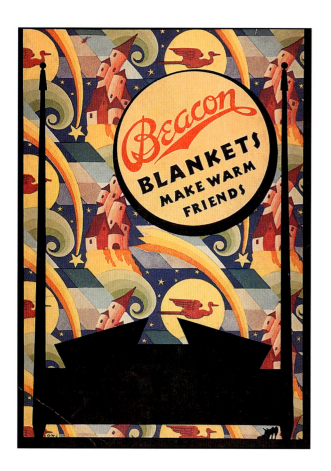

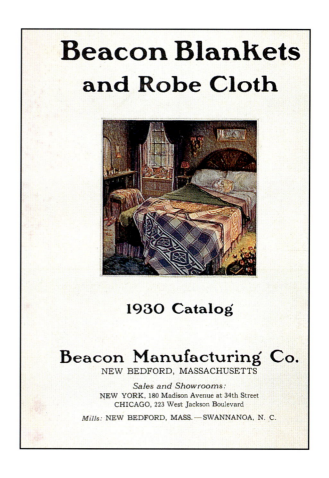

When we say Beacon Blankets make warm friends we mean that shoppers who
know value select Beacon Blankets because Beacon quality
and patterns are appreciated by them.

[2]

Ombré Bathrobe Blankets

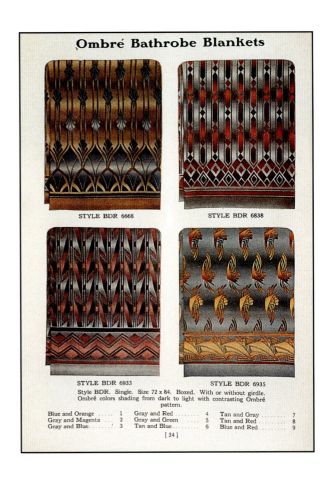

STYLE BDR 6668

STYLE BDR 6838

STYLE BDR 6933

STYLE BDR 6935

Style BDR. Single. Size 72 x 84. Boxed. With or without girdle.
Ombré colors shading from dark to light with contrasting Ombré
pattern.

Blue and Orange	1	Gray and Red	4	Tan and Gray	7
Gray and Magenta	2	Gray and Green	5	Tan and Red	8
Gray and Blue	3	Tan and Blue	6	Blue and Red	9

[34]

Signet Blankets

Blue, Tan, Buff 14   Rose and Buff 91   Blue and White 14W

STYLE SIGNET 3

Orchid and White 70W   Rose and White 91W   Orchid, Tan, Buff 70

STYLE SIGNET 7

Style SIGNET. Single. Size 66 x 80. Ends bound with 3-inch sateen.

Blue, Tan, Buff	14	Orchid, Tan, Buff	70	Brown, Taupe, Tan	18
Blue and White	14W	Orchid and White	70W	Copenhagen, Buff, Blue	55
Rose, Buff	91	Green and Buff	86	Maroon, Taupe, Buff	78
Rose and White	91W	Gold and White	19W	Navy, Taupe, Buff	23

[14]

Ombré Robe Cloth

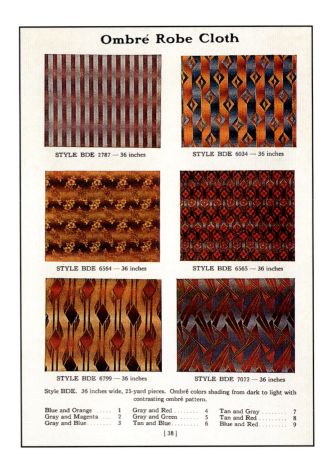

STYLE BDE 2787 — 36 inches

STYLE BDE 6034 — 36 inches

STYLE BDE 6564 — 36 inches

STYLE BDE 6565 — 36 inches

STYLE BDE 6799 — 36 inches

STYLE BDE 7072 — 36 inches

Style BDE. 36 inches wide, 25-yard pieces. Ombré colors shading from dark to light with
contrasting ombré pattern.

Blue and Orange	1	Gray and Red	4	Tan and Gray	7
Gray and Magenta	2	Gray and Green	5	Tan and Red	8
Gray and Blue	3	Tan and Blue	6	Blue and Red	9

[38]

Sample Case No. 1

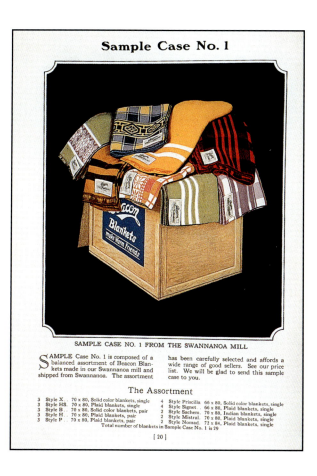

SAMPLE CASE NO. 1 FROM THE SWANNANOA MILL

SAMPLE Case No. 1 is composed of a balanced assortment of Beacon Blankets made in our Swannanoa mill and shipped from Swannanoa. The assortment has been carefully selected and affords a wide range of good sellers. See our price list. We will be glad to send this sample case to you.

The Assortment

3	Style X . .	70 x 80, Solid color blankets, single
3	Style HS .	70 x 80, Plaid blankets, single
3	Style B . .	70 x 80, Solid color blankets, pair
3	Style H . .	70 x 80, Plaid blankets, pair
3	Style P . .	70 x 80, Plaid blankets, pair
4	Style Priscilla	66 x 80, Solid color blankets, single
4	Style Signet .	66 x 80, Plaid blankets, single
2	Style Sachem .	70 x 80, Indian blankets, single
2	Style Mistral .	70 x 80, Plaid blankets, single
2	Style Nomad .	72 x 84, Plaid blankets, single

Total number of blankets in Sample Case No. 1 is 29

[20]

Agawam Blankets

STYLE AGAWAM 28   STYLE AGAWAM 37

Style AGAWAM 28 and 37. Single. Size 60 x 80. Bound around with 1½ inch tape. Boxed.

STYLE 28. COLORS		STYLE 37. COLORS	
Navy, Peacock, Taupe, Orange	23	Fawn, Slate, Peacock, Red	1
Fawn, Red, Black, Peacock	44	Fawn, Blue, Green, Red	2
Tan, Orange, Green, Purple	25	Fawn, Red, Peacock, Green	3
Green, Gold, Red, Peacock	90	Taupe, Red, Blue, Orange	4
Light Blue, Red, Black, Gold	14	Fawn, Brown, Peacock, Red	5
Slate, Purple, Black, Orange	48	Tan, Orange, Red, Blue, Green	6
Red, Gold, Green, Peacock	45	Slate, Green, Lavender, Brown	7
		Tan, Orchid, Green, Orange	8
		Taupe, Green, Orchid, Orange	9
		Light Brown, Green, Red, Blue	10

[31]

Sample Case No. 2

SAMPLE CASE NO. 2 FROM THE NEW BEDFORD MILL

A wide range of style leaders and good sellers.
See our price list.

The Assortment

3	Style Yukon	70 x 80, Plaid blankets, single
2	Style PY .	70 x 80, Comfortables, single
2	Style CO .	70 x 80, Ombré blankets, single
2	Style PR .	72 x 84, Comfortables, single
1	Style PC . .	72 x 84, Comfortables, single
4	Style Agawam	60 x 80, Indian blankets, single
2	Style Casco .	60 x 80, Ombré Indian blankets, single
2	Style TR . .	60 x 80, Ombré Plain Robes, single

Clippings of Robe Cloth—Styles E, EB, and BDE also included
Total number of blankets in Sample Case No. 2 is 18

[39]

Agawam Blankets

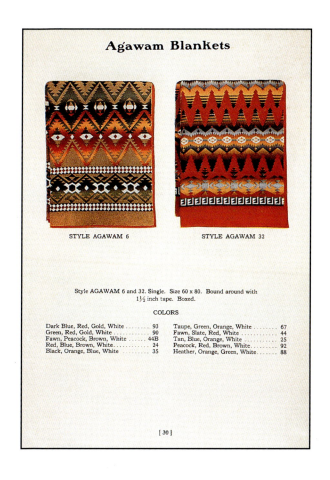

STYLE AGAWAM 6   STYLE AGAWAM 32

Style AGAWAM 6 and 32. Single. Size 60 x 80. Bound around with 1½ inch tape. Boxed.

COLORS

Dark Blue, Red, Gold, White	93	Taupe, Green, Orange, White	67
Green, Red, Gold, White	90	Fawn, Slate, Red, White	44
Fawn, Peacock, Brown, White	44B	Tan, Blue, Orange, White	25
Red, Blue, Brown, White	24	Peacock, Red, Brown, White	92
Black, Orange, Blue, White	35	Heather, Orange, Green, White	88

[30]

Sachem Blankets

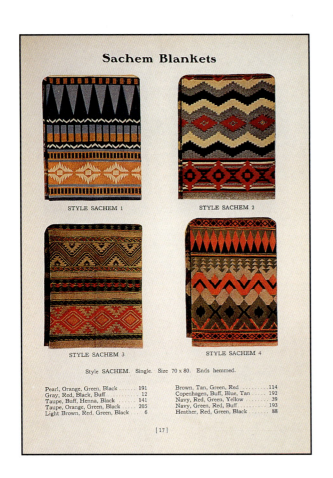

STYLE SACHEM 1   STYLE SACHEM 2

STYLE SACHEM 3   STYLE SACHEM 4

Style SACHEM. Single. Size 70 x 80. Ends hemmed.

Pearl, Orange, Green, Black	191	Brown, Tan, Green, Red	114
Gray, Red, Black, Buff	12	Copenhagen, Buff, Blue, Tan	192
Taupe, Buff, Henna, Black	141	Navy, Red, Green, Yellow	39
Taupe, Orange, Green, Black	205	Navy, Green, Red, Buff	193
Light Brown, Red, Green, Black	6	Heather, Red, Green, Black	88

[17]

Casco Blankets

STYLE CASCO 1   STYLE CASCO 3

Style CASCO. Single. Size 60 x 80. Bound all around with 1½-inch tape. Boxed.

STYLE CASCO 1 ONLY.
COLORS

Tan, White, Orange, Red	7	Heather, White, Green, Orange	10
Taupe, White, Green, Black	1	Peacock, White, Orange, Red	8
Slate, White, Red, Maroon	4	Red, White, Green, Black	6
Purple, White, Green, Black	5		

Casco 1 is the same pattern as Casco 3, but has more White in the body.

STYLE CASCO 3.
COLORS

Pearl, Orange, Red, Green, White, Black	6	Green, Orange, Yellow, Red, White, Maroon	4
Taupe, Green, Orange, Red, White, Black	3	Dark Blue, Red, Maroon, Yellow, Orange, White	7
Turquoise, Yellow, Orange, Black, Green, White	10	Navy, Orange, Red, Green, White, Gray	5
Red, Green, Black, Brown, Tan, White	12	Heather, Green, Orange, Red, White	14

[32]

Casco Blankets

STYLE CASCO 10   STYLE CASCO 12

Style CASCO. Single. Size 60 x 80. Bound all around with 1½ inch tape. Boxed.

CASCO 10-12-17-19

Colors same as Style Casco 3, described on page 32.

STYLE CASCO 17   STYLE CASCO 19

[33]

Ombré Plaid Robe

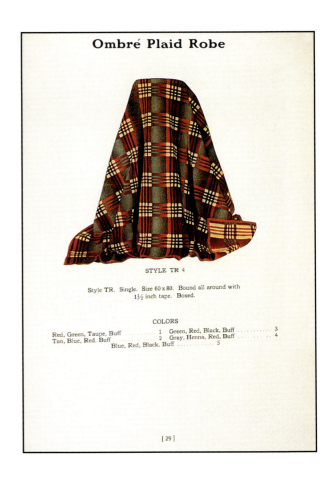

STYLE TR 4

Style TR. Single. Size 60 x 80. Bound all around with 1½ inch tape. Boxed.

COLORS

Red, Green, Taupe, Buff	1	Green, Red, Black, Buff	3
Tan, Blue, Red, Buff	2	Gray, Henna, Red, Buff	4
Blue, Red, Black, Buff	5		

[29]

80

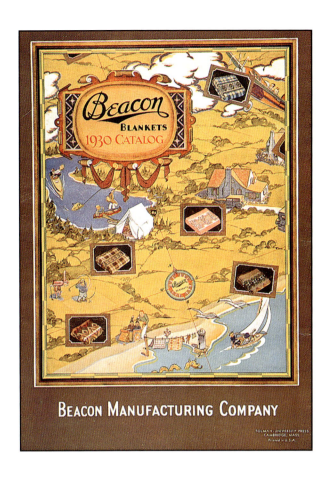

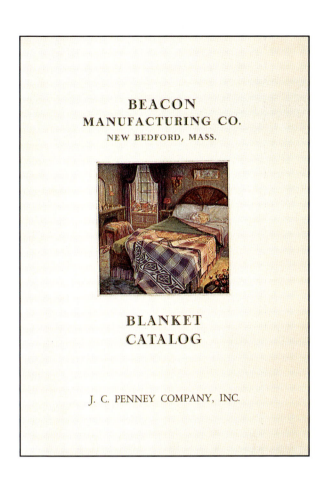

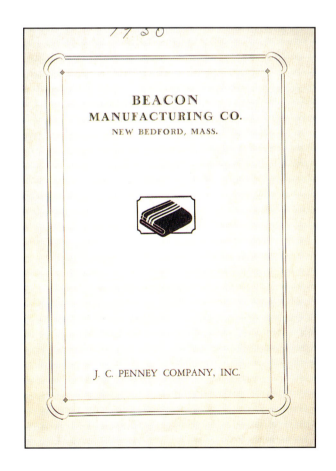

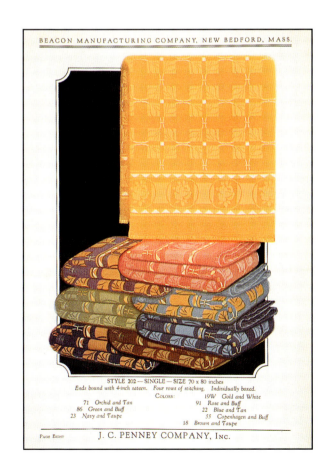

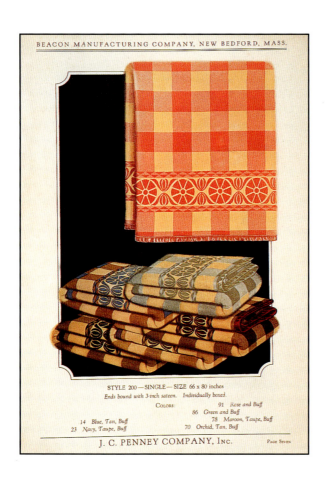

BEACON MANUFACTURING COMPANY, NEW BEDFORD, MASS.

STYLE 200—SINGLE—SIZE 66 x 80 inches
Ends bound with 3-inch sateen.   Individually boxed.
Colors:
91 Rose and Buff
86 Green and Buff
14 Blue, Tan, Buff   78 Maroon, Taupe, Buff
23 Navy, Taupe, Buff   70 Orchid, Tan, Buff

J. C. PENNEY COMPANY, Inc.   Page Seven

Courtesy of Gary Holt/Steve Christianson, Gary Holt Collection.

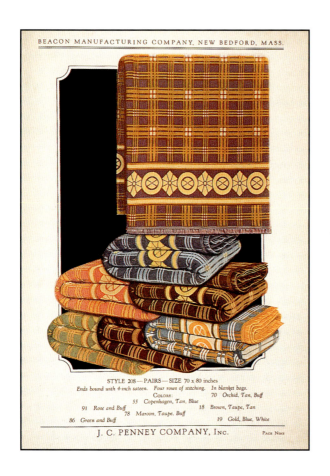

BEACON MANUFACTURING COMPANY, NEW BEDFORD, MASS.

STYLE 208—PAIRS—SIZE 70 x 80 inches
Ends bound with 4-inch sateen.   Four rows of stitching.   In blanket bags.
Colors: 70 Orchid, Tan, Buff
55 Copenhagen, Tan, Blue
91 Rose and Buff   18 Brown, Taupe, Tan
78 Maroon, Taupe, Buff
86 Green and Buff   19 Gold, Blue, White

J. C. PENNEY COMPANY, Inc.   Page Nine

Author's collection. $50-200.

A well-appreciated Ombre blanket that has spent time out on the range, "c. 1927." $75-125. Author's collection.

An Ombre shawl produced in 1925/1926. Excellent design with an orange fringe to match. Courtesy of Bearwallow Mountain Traders, Judy Hudson & Norwood Barnes.

Sensational color combination and design layout, late-1920s. Courtesy of Bearwallow Mountain Traders, Judy Hudson & Norwood Barnes.

Beacon blankets galore. Courtesy of Sam and Denise Kennedy of Cisco's, Coeur d'Alene, Idaho.

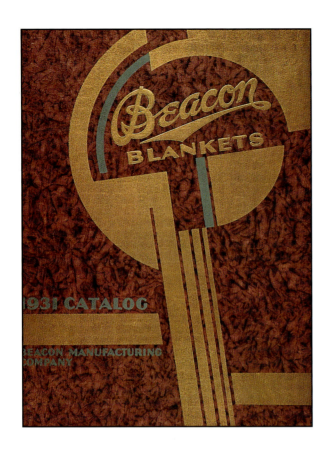

SIGNET BLANKETS—Single

Style Signet 58

Style Signet 59

STYLE SIGNET. Single. Size 66 x 80. Ends bound with 3-inch sateen.

COLOR COMBINATIONS FOR ALL SIGNETS

Blue, Tan, and Buff	14	Brown, Taupe, and Tan.... 18	Tan, Taupe, and White 67
Rose, Tan, and Buff	91	Copenhagen, Buff, and Blue 55	Orchid and White...... 70W
Green and Buff	86	Maroon, Taupe, and Buff.. 78	Blue and White........ 14W
Orchid, Tan, and Buff	70	Navy, Taupe, and Buff..... 23	Rose and White........ 91W
Green and White.....86W		Gold and White 19W made only in Style Signet 14	

THE blankets shown thus far in this catalog were made in our ideally located Swannanoa Mill in North Carolina. We wish to call special attention to new Swannanoa numbers shown for the first time in our line.

LARK — a single blanket, size 66 x 80, in plaid designs and coloring. (Page 5)

WHALE — an extra large blanket, size 80 x 90. Ends bound with 4-inch sateen. 4 rows of stitching. This is a 5-inch block design. (Page 8)

IRIS — a single blanket. Size 70 x 80. Ends bound with 4-inch sateen in both dark and light color combinations. (Pages 18 and 19)

All these blankets are shipped F. O. B. Swannanoa, N. C.

LARK PLAID BLANKETS—Single

Style Lark 1

Style Lark 2

Style Lark 3

STYLE LARK. Single. Size 66 x 80. Ends bound with 3-inch sateen. Fancy Jacquard designs.

COLORS

Blue, Light Blue, Buff	22	Brown, Taupe, Tan	18
Rose, Light Rose, Buff	29	Copenhagen, Tan, Blue	55
Orchid, Light Orchid, Buff	71	Green, Light Green, Buff	86

NOMAD BLANKETS—Single

Style Nomad 4   Style Nomad 5

Style Nomad 6

STYLE NOMAD. Single. Size 72 x 84. Ends bound with 4-inch sateen.
4 rows of stitching

COLOR COMBINATIONS FOR ALL NOMAD

Blue, Tan, and Buff....	14	Brown, Taupe, and Tan....	18	Tan, Taupe, and White	67
Rose, Tan, and Buff..	91	Copenhagen, Buff, and Blue	55	Orchid and White......	70W
Green and Buff......	86	Maroon, Taupe, and Buff..	78	Blue and White.......	14W
Orchid, Tan, and Buff	70	Navy, Taupe, and Buff	23	Rose and White........	91W
Green and White.....86W		Gold and White 19W made only in Style Nomad 5			

Page 17

IRIS BLANKETS—Single

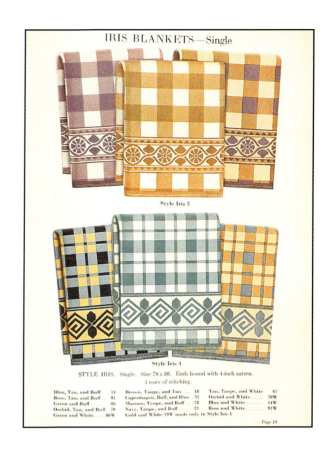

Style Iris 3

Style Iris 4

STYLE IRIS. Single. Size 70 x 80. Ends bound with 4-inch sateen.
4 rows of stitching.

Blue, Tan, and Buff....	14	Brown, Taupe, and Tan....	18	Tan, Taupe, and White	67
Rose, Tan, and Buff..	91	Copenhagen, Buff, and Blue	55	Orchid and White......	70W
Green and Buff......	86	Maroon, Taupe, and Buff..	78	Blue and White.......	14W
Orchid, Tan, and Buff	70	Navy, Taupe, and Buff	23	Rose and White........	91W
Green and White.....86W		Gold and White 19W made only in Style Iris 1			

Page 19

PLAID BLANKETS—Pairs

Style P 5   For colors of P 5 see page 11   Style P 117

Style P 119   Style P 121

STYLE P. Pairs. Size 70 x 80. Ends bound with 4-inch sateen. 4 rows
of stitching. Fancy Jacquard designs.

COLORS: STYLE P 117–P 119–P 121.

Blue, Light Blue, Buff............	22	Brown, Taupe, Tan................	18
Rose, Light Rose, Buff............	29	Copenhagen, Tan, Blue............	55
Orchid, Light Orchid, Buff........	71	Green, Light Green, Buff..........	86

Page 10

OMBRÉ ROBE CLOTH

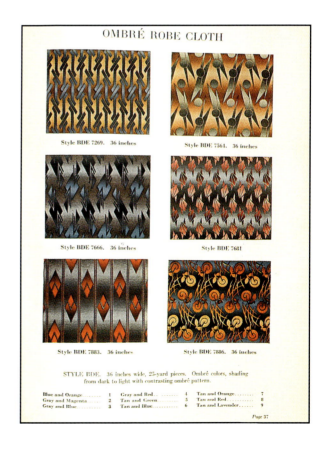

Style BDE 7269.   36 inches   Style BDE 7561.   36 inches

Style BDE 7666.   36 inches   Style BDE 7681

Style BDE 7883.   36 inches   Style BDE 7886.   36 inches

STYLE BDE. 36 inches wide, 25-yard pieces. Ombré colors, shading
from dark to light with contrasting ombré pattern.

Blue and Orange.........	1	Gray and Red.........	4	Tan and Orange........	7
Gray and Magenta........	2	Tan and Green.........	5	Tan and Red..........	8
Gray and Blue..........	3	Tan and Blue...........	6	Tan and Lavender.....	9

Page 37

OMBRÉ BLANKETS

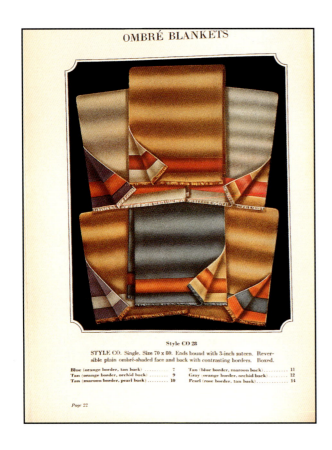

Style CO 28

STYLE CO. Single. Size 70 x 80. Ends bound with 3-inch sateen. Reversible plain ombré-shaded face and back with contrasting borders. Boxed.

Blue (orange border, tan back)	7	Tan (blue border, maroon back)	11
Tan (orange border, orchid back)	9	Gray (orange border, orchid back)	12
Tan (maroon border, pearl back)	10	Pearl (rose border, tan back)	14

OMBRÉ BLANKETS—Single

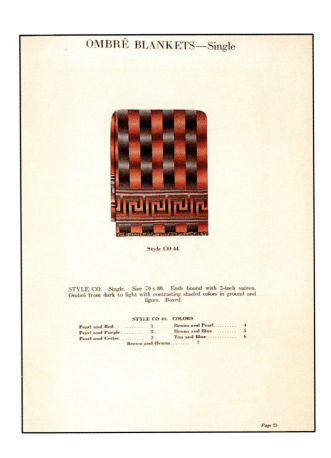

Style CO 44

STYLE CO. Single. Size 70 x 80. Ends bound with 3-inch sateen. Ombré from dark to light with contrasting shaded colors in ground and figure. Boxed.

STYLE CO 44. COLORS

Pearl and Red	1	Henna and Pearl	4
Pearl and Purple	2	Henna and Blue	5
Pearl and Cerise	3	Tan and Blue	6
	Brown and Henna	7	

AGAWAM BLANKETS—Single

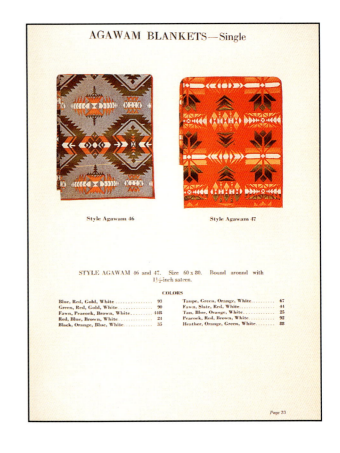

Style Agawam 46	Style Agawam 47

STYLE AGAWAM 46 and 47. Size 60 x 80. Bound around with 1½-inch sateen.

COLORS

Blue, Red, Gold, White	93	Taupe, Green, Orange, White	67
Green, Red, Gold, White	90	Fawn, Slate, Red, White	44
Fawn, Peacock, Brown, White	44B	Tan, Blue, Orange, White	25
Red, Blue, Brown, White	24	Peacock, Red, Brown, White	92
Black, Orange, Blue, White	35	Heather, Orange, Green, White	88

AGAWAM BLANKETS—Single

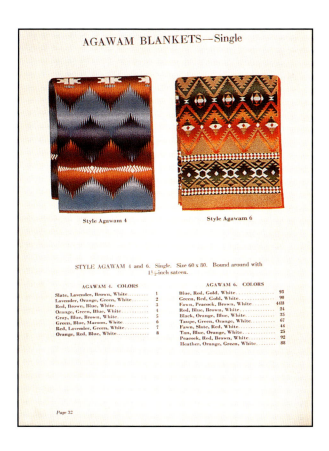

Style Agawam 4   Style Agawam 6

STYLE AGAWAM 4 and 6. Single. Size 60 x 80. Bound around with 1½-inch sateen.

AGAWAM 4. COLORS		AGAWAM 6. COLORS	
Slate, Lavender, Brown, White	1	Blue, Red, Gold, White	93
Lavender, Orange, Green, White	2	Green, Red, Gold, White	90
Red, Brown, Blue, White	3	Fawn, Peacock, Brown, White	44B
Orange, Green, Blue, White	4	Red, Blue, Brown, White	24
Gray, Blue, Brown, White	5	Black, Orange, Blue, White	35
Green, Blue, Maroon, White	6	Taupe, Green, Orange, White	67
Red, Lavender, Green, White	7	Fawn, Slate, Red, White	44
Orange, Red, Blue, White	8	Tan, Blue, Orange, White	25
		Peacock, Red, Brown, White	92
		Heather, Orange, Green, White	88

Page 32

WIGWAM BLANKETS—Single

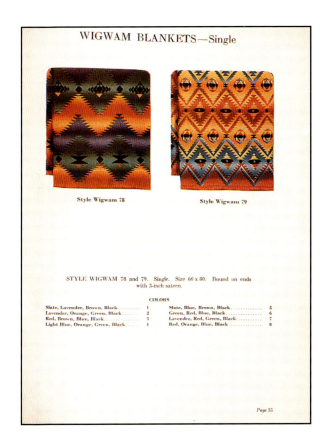

Style Wigwam 78   Style Wigwam 79

STYLE WIGWAM 78 and 79. Single. Size 66 x 80. Bound on ends with 3-inch sateen.

COLORS			
Slate, Lavender, Brown, Black	1	Slate, Blue, Brown, Black	5
Lavender, Orange, Green, Black	2	Green, Red, Blue, Black	6
Red, Brown, Blue, Black	3	Lavender, Red, Green, Black	7
Light Blue, Orange, Green, Black	4	Red, Orange, Blue, Black	8

Page 35

WIGWAM BLANKETS—Single

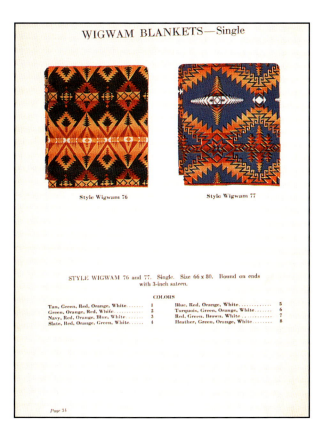

Style Wigwam 76   Style Wigwam 77

STYLE WIGWAM 76 and 77. Single. Size 66 x 80. Bound on ends with 3-inch sateen.

COLORS			
Tan, Green, Red, Orange, White	1	Blue, Red, Orange, White	5
Green, Orange, Red, White	2	Turquois, Green, Orange, White	6
Navy, Red, Orange, Blue, White	3	Red, Green, Brown, White	7
Slate, Red, Orange, Green, White	4	Heather, Green, Orange, White	8

Page 34

Diamonds, snowflakes and Ombre shading create a kaleidoscopic look on a three color blanket, late-1920s. Courtesy of Bearwallow Mountain Traders, Judy Hudson & Norwood Barnes.

Several motifs converge on this Ombre blanket bound on all four sides, late-1920s. Courtesy of Bearwallow Mountain Traders, Judy Hudson & Norwood Barnes.

Outstanding Indian Ombre design in perfect condition, late-1920. Courtesy of Bearwallow Mountain Traders, Judy Hudson & Norwood Barnes.

One of Cisco's many displays of beautiful Beacon blankets. Courtesy of Sam and Denise Kennedy of Cisco's, Coeur d'Alene, Idaho.

An Indian design Ombre bound on all four sides, late-1920s. Courtesy of Bearwallow Mountain Traders, Judy Hudson & Norwood Barnes.

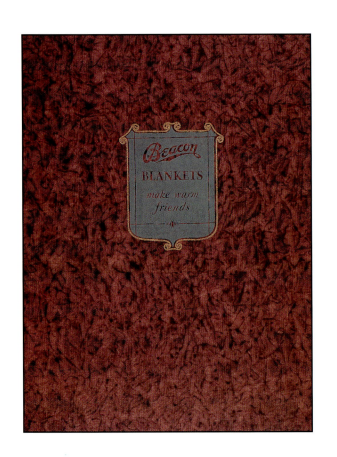

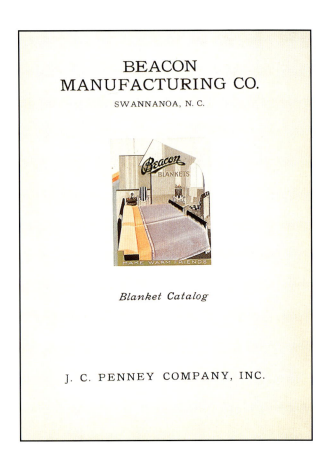

BEACON
MANUFACTURING CO.

SWANNANOA, N. C.

Blanket Catalog

J. C. PENNEY COMPANY, INC.

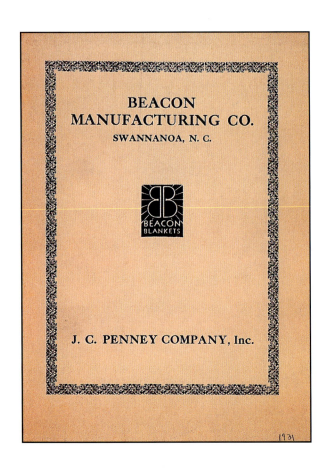

BEACON
MANUFACTURING CO.

SWANNANOA, N. C.

BEACON
BLANKETS

J. C. PENNEY COMPANY, Inc.

1931

STYLE 214 — SINGLE.   *Size 66 x 80 inches.*
Ends bound with 3-inch sateen.
In blanket bags.

COLORS

| 12. Gray, Red, Black and Buff | 191. Pearl, Orange, Green and Black |
| 39. Navy, Red, Green and Yellow | 141. Taupe, Buff, Henna, and Black |

J. C. PENNEY COMPANY, Inc.   *Page 3*

90

STYLE 212 — SINGLE. *Size 66 x 80 inches.*
Ends bound with 3-inch sateen.
In blanket bags.
COLORS
86. Green and Buff   23. Navy, Taupe and Buff
70. Orchid, Tan and Buff   78. Maroon, Taupe and Buff

J. C. PENNEY COMPANY, Inc.   Page 5

STYLE 216 — SINGLE. *Size 70 x 80 inches.*
Ends bound with 4-inch sateen. Four rows of stitching.
Individually boxed.
COLORS
39. Navy, Red, Green and Yellow   191. Pearl, Orange, Green and Black
205. Taupe, Orange, Green and Black

Page 6   J. C. PENNEY COMPANY, Inc.

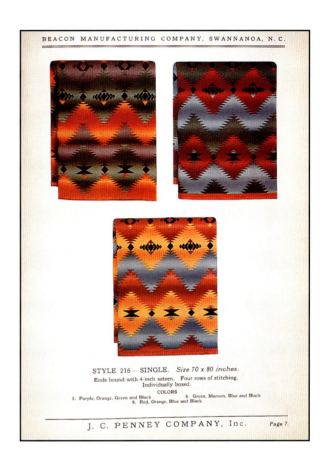

STYLE 216 — SINGLE. *Size 70 x 80 inches.*
Ends bound with 4-inch sateen. Four rows of stitching.
Individually boxed.
COLORS
2. Purple, Orange, Green and Black   6. Green, Maroon, Blue and Black
8. Red, Orange, Blue and Black

J. C. PENNEY COMPANY, Inc.   Page 7

End of An Era

In the late 1940s, Beacon started blending synthetic fibers into their fabrics. By the 1950s, the *Ombre* blankets were no longer made. The Indian designs were still in the line, as were the robe fabrics, but new blends gave a different feeling to these products. The original *Ombre* process, a trademark for Beacon products from the mid-1920s through the mid-1950s, would become a part of Beacon's history. Today, these blankets and robes are collectors' items.

Bolder Graphics: 1932-1950

In 1932 the graphic art designs became larger and more dramatic.

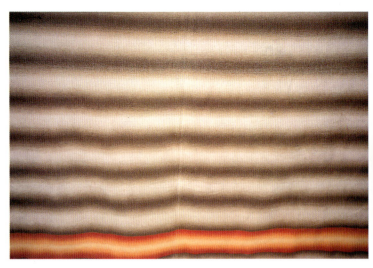

Reverse colorization side of the Ombre stripes appears to be another blanket but they are one in the same, Courtesy of Gary Holt/Steve Christianson, Gary Holt collection.

An old wooden ladder makes an interesting display prop for a collection of Beacon Ombre blankets and shawls, "c. 1930s." Courtesy of Bearwallow Mountain Traders, Judy Hudson & Norwood Barnes.

Totally vibrant Ombre shawl ready for a sleigh ride, "c. 1930." Courtesy of Bearwallow Mountain Traders, Judy Hudson & Norwood Barnes.

Ombre stripes become electric on this 1928 Beacon blanket, patternization side. Originally sold as reversible. Courtesy of Gary Holt/Steve Christianson, Gary Holt collection.

More wardrobe closets brim full of beautiful Beacon blankets. Courtesy of Sam and Denise Kennedy of Cisco's, Coeur d'Alene, Idaho.

...patternization ...m and Denise ...ene, Idaho.

Colorization side of an Inca design blanket, "c. 1932." Courtesy of Sam and Denise Kennedy of Cisco's, Coeur d'Alene, Idaho.

A variation of the Inca design scheme, patternization side, early-1930s.
Courtesy of Sam and Denise Kennedy of Cisco's, Coeur d'Alene, Idaho.

The colorization side of a variation of the Inca design, early-1930s.
Courtesy of Sam and Denise Kennedy of Cisco's, Coeur d'Alene, Idaho.

An Inca shawl, colorization side, with red fringe, "c. 1932." Courtesy of Bearwallow Mountain Traders, Judy Hudson & Norwood Barnes.

The Inca design shawl in green with green fringe, patternization side with split designs on the edges, "c. 1935/1936." Courtesy of Bearwallow Mountain Traders, Judy Hudson & Norwood Barnes.

A bold red and blue Inca design Ombre blanket, "c. 1935/1936." Courtesy of Bearwallow Mountain Traders, Judy Hudson & Norwood Barnes.

Red and tan Inca shawl with red fringe, patternization side, "c. 1932." Courtesy of Bearwallow Mountain Traders, Judy Hudson & Norwood Barnes.

The reverse side of the red Inca blanket. Reversed, the blanket color is predominately blue, "c. 1935/1936." Courtesy of Bearwallow Mountain Traders, Judy Hudson & Norwood Barnes.

THE 1932 BEACON LINE

● The Blankets shown thus far in this catalog were made in our southern mill at Swannanoa, North Carolina. Among the many patterns and numbers are the new Dalton (page 8), Cameo (page 5), and Kismet blankets (pages 14-15-16) which deserve special mention. All blankets in this group are shipped F. O. B. Swannanoa, N. C.

● Part 2 contains blankets from the New Bedford mill only. The outstanding new numbers are the large ombre plaid designs in the Yukon group (pages 26-27) and the Inca Indian designs with large overall ombre patterns (pages 36-37). All blankets in this group are shipped F. O. B. New Bedford, Mass.

● Part 3 of this catalog shows the full Beacon line of Crib Blankets, including many new patterns and styles never before shown. The price range is from the inexpensive Style S line to the more expensive Style F and LG blankets and the Fringed Carriage Robes. All Crib Blankets are made in our mill at New Bedford and shipped F. O. B. mill, New Bedford, Mass.

● Beacon Blankets are on display at our sales and show rooms, 180 Madison Avenue, New York, and 223 West Jackson Boulevard, Chicago. Executive offices are in New Bedford, Mass.

24

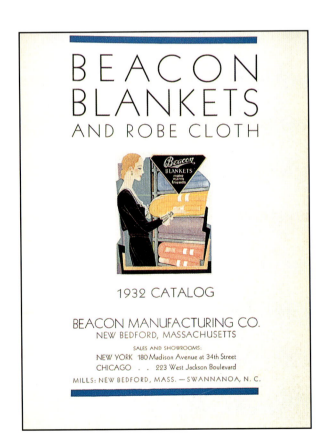

SIGNET BLANKETS—Single

STYLE SIGNET 16

STYLE SIGNET 17

STYLE SIGNET. Single. Size 70 x 80. Ends bound with 4-inch sateen. 4 rows of stitching.

COLOR COMBINATIONS FOR ALL SIGNETS

Blue, Tan, and Buff . 14	Green and White . 86W	Brown, Taupe, and Tan 18	Tan, Taupe, and White 67
Rose, Tan, and Buff . 91	Orchid, Tan, and Buff 70	Maroon, Taupe, and .	Orchid and White 70W
Green and Buff . 86	Copenhagen, Buff,	Buff 78	Blue and White . 14W
Green, Tan, Buff . . 86T	and Blue 55	Navy, Taupe, and Buff 23	Rose and White . 91W
	Gold and White 19W (made only in Style Signet 16)		

17

96

SIGNET BLANKETS—Single

STYLE SIGNET 58

STYLE SIGNET 59

STYLE SIGNET. Single. Size 70 x 80. Ends bound with 4-inch sateen. 4 rows of stitching.

COLOR COMBINATIONS FOR ALL SIGNETS

Blue, Tan, and Buff	. 14	Green and White	. 86W	Brown, Taupe, Tan	. 18	Tan, Taupe, White . . 67 -
Rose, Tan, and Buff	. 91	Orchid, Tan, Buff	. 70	Maroon, Taupe, and		Orchid and White . 70W
Green and Buff	. 86	Copenhagen, Buff,		Buff	. 78	Blue and White . 14W
Green, Tan, Buff	. 86T	Blue	. 55	Navy, Taupe, Buff . 23		Rose and White . 91W

Gold and White 19W (made only in Style Signet 16)

18

NOMAD BLANKETS—Single

STYLE NOMAD 10

STYLE NOMAD 11

STYLE NOMAD. Single. Size 72 x 84. Ends bound with 4-inch sateen. 4 rows of stitching.

COLOR COMBINATIONS FOR ALL NOMADS

Blue, Tan, and Buff	. 14	Orchid, Tan, Buff	. 70	Brown, Taupe, Tan	. 18	Tan, Taupe, White . . 67 -
Rose, Tan, and Buff	. 91	Green and White	. 86W	Maroon, Taupe, and		Orchid and White 70W
Green and Buff	. 86	Copenhagen, Buff,		Buff	. 78	Blue and White . 14W
Green, Tan, Buff	. 86T	Blue	. 55	Navy, Taupe, Buff . 23		Rose and White . 91W

Gold and White 19W (made only in Style Nomad 5)

21

NOMAD BLANKETS—Single

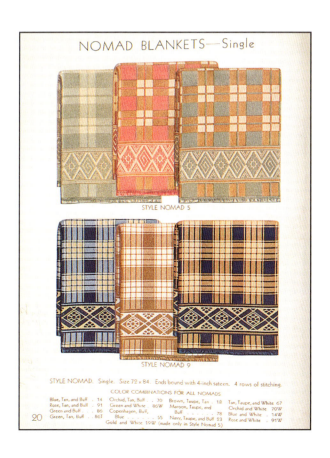

STYLE NOMAD 5

STYLE NOMAD 9

STYLE NOMAD. Single. Size 72 x 84. Ends bound with 4-inch sateen. 4 rows of stitching.

COLOR COMBINATIONS FOR ALL NOMADS

Blue, Tan, and Buff	. 14	Orchid, Tan, Buff	. 70	Brown, Taupe, Tan	. 18	Tan, Taupe, and White 67
Rose, Tan, and Buff	. 91	Green and White	. 86W	Maroon, Taupe, and		Orchid and White 70W
Green and Buff	. 86	Copenhagen, Buff,		Buff	. 78	Blue and White . 14W
Green, Tan, Buff	. 86T	Blue	. 55	Navy, Taupe, and Buff 23		Rose and White . 91W

Gold and White 19W (made only in Style Nomad 5)

20

KISMET PLAID BLANKETS—Single

STYLE KISMET 22

STYLE KISMET 19

STYLE KISMET. Single. Size 66 x 80. Ends bound with 3-inch sateen.

COLOR COMBINATIONS FOR ALL KISMET

Blue, Tan, and Buff	. 14	Brown, Taupe, and Tan	. . . 18	Tan, Taupe, and White . 67
Rose, Tan, and Buff	. 91	Copenhagen, Buff, and Blue	. 55	Orchid and White . 70W
Green and Buff	. 86	Maroon, Taupe, and Buff	. 78	Blue and White . 14W
Orchid, Tan, and Buff	. 70	Navy, Taupe, and Buff . 23		Rose and White . 91W
Green and White	. 86W			Green, Tan, Buff . . 86T

Gold and White 19W (made only in Style Kismet 22)

14

KISMET PLAID BLANKETS—Single

STYLE KISMET 20

STYLE KISMET 18

STYLE KISMET. Single. Size 66 x 80. Ends bound with 3-inch sateen.

COLOR COMBINATIONS FOR ALL KISMET

Blue, Tan, and Buff . . 14	Brown, Taupe, and Tan . . . 18	Tan, Taupe, and White . 67
Rose, Tan, and Buff . . 91	Copenhagen, Buff, and Blue 55	Orchid and White . . 70W
Green and Buff 86	Maroon, Taupe, and Buff . . 78	Rose and White . . . 91W
Orchid, Tan and Buff . . 70	Navy, Taupe, and Buff . . 23	Blue and White . . . 14W
Green and White . . . 86W		Green, Tan, Buff . . 86T
	Gold and White 19W (made only in Style Kismet 22)	

15

PLAID BLANKETS—Pairs

STYLE P 70

STYLE P 118

STYLE P 70—P 118
COLORS

Gold, White, Green	19
Blue, White, Tan . .	22
Rose, White, Tan . .	29
Tan, White, Blue . .	44
Orchid, White, Tan .	71
Green, White, Gold	83

STYLE P 126
COLORS

Brown and Buff . .	18
Blue and Buff . . .	23
Lavender and Buff .	51
Red and Buff	78
Green and Buff . .	90

STYLE P 126

STYLE P. Pairs. Size 72 x 84. Ends bound with 4-inch sateen. 4 rows of stitching. Fancy Jacquard designs.

11

PLAID BLANKETS—Pairs

STYLE P 5

STYLE P 117

STYLE P 5
COLORS

Brown and Tan . .	18
Gold and White . .	19
Blue and White . .	22
Tan and White . .	25
Rose and White . .	29
Red and Black . .	35
Orchid and White .	71
Green and White .	86
Pearl and White . .	89

STYLE P 117—P 121
COLORS

Blue, Tan, Buff . . .	14
Brown, Taupe, Tan .	18
Blue, Light Blue, Buff	22
Navy, Taupe, Buff .	23
Rose, Light Rose, Buff	29
Copenhagen, Tan, Blue	55
Orchid, Tan, Buff . .	70
Orchid, Light Orchid, Buff	71
Maroon, Taupe, Buff	78
Green, Light Green, Buff	86
Green, Tan, Buff . .	86T
Rose, Tan, Buff . . .	91

STYLE P 121

STYLE P. Pairs. Size 72 x 84. Ends bound with 4-inch sateen. 4 rows of stitching. Fancy Jacquard designs.

10

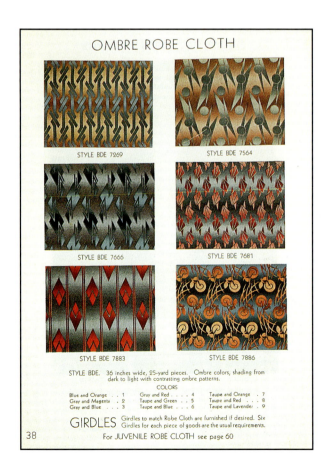

OMBRE ROBE CLOTH

STYLE BDE 7269

STYLE BDE 7564

STYLE BDE 7666

STYLE BDE 7681

STYLE BDE 7883

STYLE BDE 7886

STYLE BDE. 36 inches wide, 25-yard pieces. Ombre colors, shading from dark to light with contrasting ombre patterns.

COLORS

Blue and Orange . . 1	Gray and Red 4	Taupe and Orange . 7
Gray and Magenta . 2	Taupe and Green . . 5	Taupe and Red . . . 8
Gray and Blue . . . 3	Taupe and Blue . . . 6	Taupe and Lavender . 9

GIRDLES Girdles to match Robe Cloth are furnished if desired. Six Girdles for each piece of goods are the usual requirements.

For JUVENILE ROBE CLOTH see page 60

38

98

OMBRE PLAID ROBE and SHAWL

STYLE OPAL 1. Single. Size 60 x 80.
Bound all around with 1½-inch sateen.

STYLE OPAL 1

COLORS
STYLE OPAL 1 and OPAL SHAWL 1

Blue Ombre with Fawn	1
Red Ombre with Peacock	2
Red Ombre with Green	3
Blue Ombre with Red	4
Brown Ombre with Green	5
Red Ombre with Fawn	6
Green Ombre with Fawn	7
Golden Brown Ombre with Fawn	10
Red Ombre with Blue	11
Green Ombre with Red	12

STYLE OPAL SHAWL 1. Single. Size
60 x 80. Fringed all around with genuine
wool fringe.

STYLE OPAL SHAWL 1

31

YUKON BLANKETS—Single

STYLE YUKON 50

STYLE YUKON 51

STYLE YUKON. Single. Size 72 x 84. Ends bound with 4-inch sateen. 4 rows of
stitching. Combining the convenience of a single blanket with the warmth of a pair.

COLORS

Rose, Taupe Ombre with Dark Red	1	Red, Taupe Ombre with Dark Red	6
Golden Brown, Taupe Ombre with Dark Brown	2	Lavender, Green Ombre with Purple	7
		Rose, Gray Ombre with Dark Red	8
Orchid, Taupe Ombre with Purple	3	Orchid, Gray Ombre with Purple	9
Green, Taupe Ombre, Dark Green	4	Red, Gray Ombre with Black	10
Blue, Taupe Ombre with Dark Blue	5	Green, Taupe Ombre with Brown	11

26

YUKON BLANKETS—Single

STYLE YUKON 52

STYLE YUKON 53

STYLE YUKON. Single. Size 72 x 84. Ends bound with 4-inch sateen. 4 rows of
stitching. Combining the convenience of a single blanket with the warmth of a pair.

COLORS

Rose, Taupe Ombre with Dark Red	1	Red, Taupe Ombre with Dark Red	6
Golden Brown, Taupe Ombre with Dark Brown	2	Lavender, Green Ombre with Purple	7
		Rose, Gray Ombre with Dark Red	8
Orchid, Taupe Ombre with Purple	3	Orchid, Gray Ombre with Purple	9
Green, Taupe Ombre, Dark Green	4	Red, Gray Ombre with Black	10
Blue, Taupe Ombre with Dark Blue	5	Green, Taupe Ombre with Brown	11

27

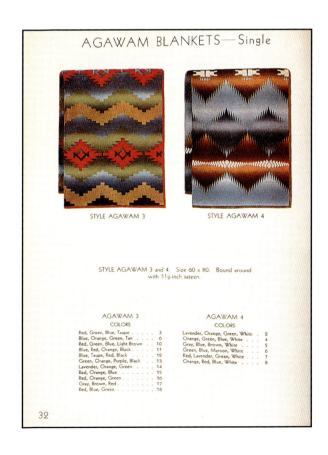

AGAWAM BLANKETS—Single

STYLE AGAWAM 3

STYLE AGAWAM 4

STYLE AGAWAM 3 and 4. Size 60 x 80. Bound around
with 1½-inch sateen.

AGAWAM 3
COLORS

Red, Green, Blue, Taupe	3
Blue, Orange, Green, Tan	6
Red, Green, Blue, Light Brown	10
Blue, Red, Orange, Black	11
Blue, Taupe, Red, Black	12
Green, Orange, Purple, Black	13
Lavender, Orange, Green	14
Red, Orange, Blue	15
Red, Orange, Green	16
Gray, Brown, Red	17
Red, Blue, Green	18

AGAWAM 4
COLORS

Lavender, Orange, Green, White	2
Orange, Green, Blue, White	4
Gray, Blue, Brown, White	5
Green, Blue, Maroon, White	6
Red, Lavender, Green, White	7
Orange, Red, Blue, White	8

32

AGAWAM BLANKETS—Single

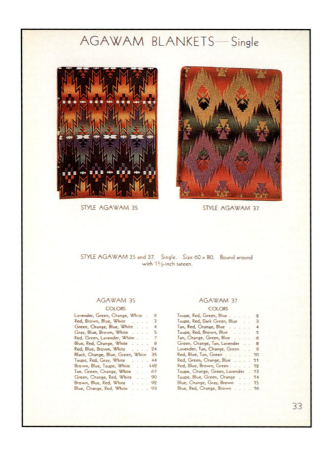

STYLE AGAWAM 35   STYLE AGAWAM 37

STYLE AGAWAM 35 and 37.   Single.   Size 60 x 80.   Bound around
with 1½-inch sateen.

AGAWAM 35 COLORS		AGAWAM 37 COLORS	
Lavender, Green, Orange, White	2	Taupe, Red, Green, Blue	2
Red, Brown, Blue, White	3	Taupe, Red, Dark Green, Blue	3
Green, Orange, Blue, White	4	Tan, Red, Orange, Blue	4
Gray, Blue, Brown, White	5	Taupe, Red, Brown, Blue	5
Red, Green, Lavender, White	7	Tan, Orange, Green, Blue	6
Blue, Red, Orange, White	8	Green, Orange, Tan, Lavender	8
Red, Blue, Brown, White	24	Lavender, Tan, Orange, Green	9
Black, Orange, Blue, Green, White	35	Red, Blue, Tan, Green	10
Taupe, Red, Gray, White	44	Red, Green, Orange, Blue	11
Brown, Blue, Taupe, White	448	Red, Blue, Brown, Green	12
Tan, Green, Orange, White	67	Taupe, Orange, Green, Lavender	13
Green, Orange, Red, White	90	Taupe, Blue, Green, Orange	14
Brown, Blue, Red, White	92	Blue, Orange, Gray, Brown	15
Blue, Orange, Red, White	93	Blue, Red, Orange, Brown	16

33

WIGWAM BLANKETS—Single

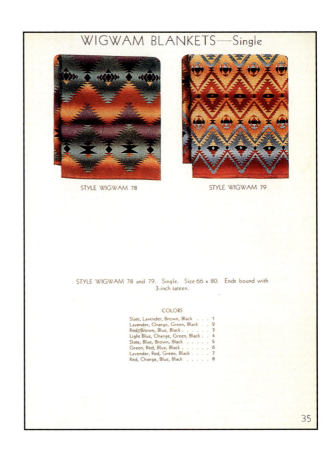

STYLE WIGWAM 78   STYLE WIGWAM 79

STYLE WIGWAM 78 and 79.   Single.   Size 66 x 80.   Ends bound with
3-inch sateen.

COLORS	
Slate, Lavender, Brown, Black	1
Lavender, Orange, Green, Black	2
Red, Brown, Blue, Black	3
Light Blue, Orange, Green, Black	4
Slate, Blue, Brown, Black	5
Green, Red, Blue, Black	6
Lavender, Red, Green, Black	7
Red, Orange, Blue, Black	8

35

WIGWAM BLANKETS—Single

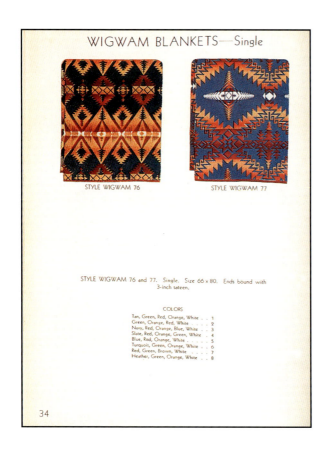

STYLE WIGWAM 76   STYLE WIGWAM 77

STYLE WIGWAM 76 and 77.   Single.   Size 66 x 80.   Ends bound with
3-inch sateen.

COLORS	
Tan, Green, Red, Orange, White	1
Green, Orange, Red, White	2
Navy, Red, Orange, Blue, White	3
Slate, Red, Orange, Green, White	4
Blue, Red, Orange, White	5
Turquois, Green, Orange, White	6
Red, Green, Brown, White	7
Heather, Green, Orange, White	8

34

INCA BLANKETS—Single

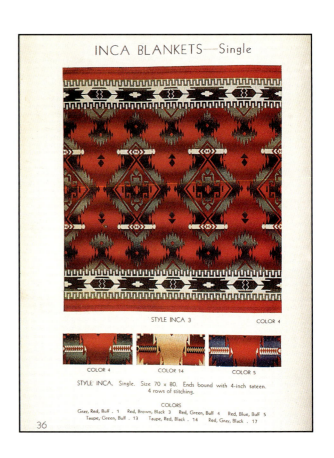

STYLE INCA 3   COLOR 4

COLOR 4   COLOR 14   COLOR 5

STYLE INCA.   Single.   Size 70 x 80.   Ends bound with 4-inch sateen.
4 rows of stitching.

COLORS

Gray, Red, Buff	1	Red, Brown, Black	3	Red, Green, Buff	4	Red, Blue, Buff	5
Taupe, Green, Buff	13	Taupe, Red, Black	14	Red, Gray, Black	17

36

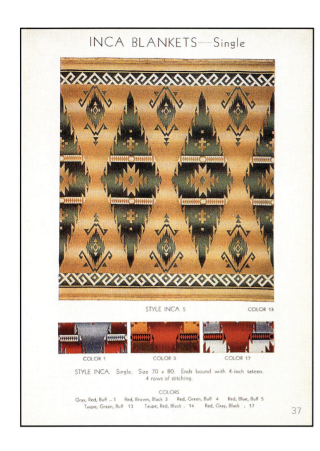

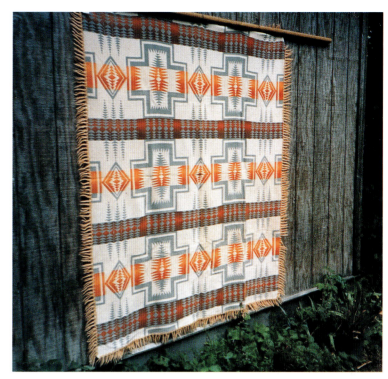

Agawam style 56, colorization side, Indian design Ombre shawl, mid-1930s. Courtesy of Bearwallow Mountain Traders, Judy Hudson & Norwood Barnes.

Agawam style 56, Indian design Ombre shawl with orange fringe, patternization side, "c. 1930s." Courtesy of Bearwallow Mountain Traders, Judy Hudson & Norwood Barnes.

Agawam style 4, Ombre shawl with a stunning color arrangement. This blanket design was available for several years, with slight variations during the 1930s. Courtesy of Bearwallow Mountain Traders, Judy Hudson & Norwood Barnes.

An interesting arrangement of plaids and diamonds with a subtle floral border and Ombre shading, "c. 1930s." Courtesy of Beacon Manufacturing Company.

Colorful Ombre Indian design, mid-1930s. Courtesy of Gary Holt/Steve Christianson, Gary Holt Collection.

Big, bold designs make a strong design statement on this mid-1930s Ombre blanket. Courtesy of Charles D. Owen III

A version of a Wigwam Ombre design from early to mid-1930s. Courtesy of Laura Fisher/Antique Quilts and Americana.

A Wigwam Ombre blanket from the early to mid-1930s in an interesting color combination. Courtesy of Charles D. Owen III.

Close up of the Beacon label on a Wigwam blanket from the early to mid-1930s. Courtesy of Charles D. Owen III.

Exceptional Wigwam style 78, "c. 1932." Courtesy Charles D. Owen III.

A well-used Indian design Ombre. Bold designs first appeared in the 1932 Beacon catalog. Courtesy of Charles D. Owen III.

Indian design Ombre blanket "c. 1930s." Courtesy of Charles D. Owen III.

Color and design combine in this exceptional Ombre blanket. "c. 1930s." Courtesy of R. Greg Otto, Portland, Oregon.

Agawam Ombre Indian design, colorization side, mid-1930s. Courtesy of Gary Holt/Steve Christianson, Gary Holt collection.

Patternization side of the Agawam Ombre, mid-1930s. Courtesy of Gary Holt/Steve Christianson, Gary Holt collection.

A good example of the large designs that appeared on Beacon blankets beginning in 1932. Courtesy of Bearwallow Mountain Traders, Judy Hudson & Norwood Barnes.

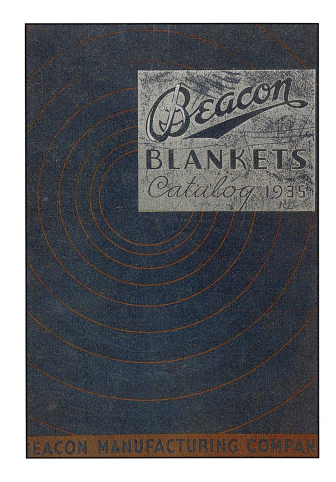

1935/1936 "Big and warm as well as colorful". Left, Indian design, 2 lbs 2 oz, 66x80, $1. Two or more, $.97 each. "Low price, many uses, no wonder it's a favorite". Center, Chippewa, 3 lbs or Vivitone, 2 lbs 11 oz, $1.59. Right, "Genuine woven Ombre shaded velvet-like beauty, vivid Indian design, large 70x80 inches" 3 lbs, $2.19. Courtesy of Sears, Roebuck & Co. archives, reprinted by special arrangement and protected by copyright. No duplication is permitted.

106

PLAID BLANKETS—Pairs

STYLE P 121

STYLE P 117

STYLE P 148

STYLE P 70

STYLE P. Pairs. Size 72 x 84. Ends bound with 4-inch sateen. 4 rows of stitching.

STYLE P 117—P 121—P 148 COLORS

		STYLE P 70 COLORS	
Blue and Tan 14	Lavender and Tan . . . 70	Gold, White and Green 19	Tan, White and Blue . . 44
Brown and Tan 18	Maroon and Tan . . . 78	Blue, White and Tan . . 22	Lavender, White and Tan 71
Navy and Tan 23	Green and Tan . . . 77	Rose, White and Tan . 29	Green, White and Gold 83
Copenhagen and Tan . 55	Rose and Tan 91		

(Page 8)

KISMET PLAID BLANKETS—Single

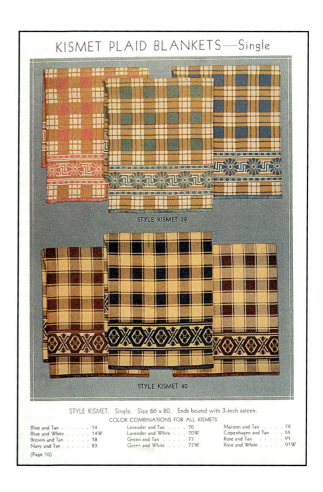

STYLE KISMET 39

STYLE KISMET 40

STYLE KISMET. Single. Size 66 x 80. Ends bound with 3-inch sateen.

COLOR COMBINATIONS FOR ALL KISMETS

Blue and Tan 14	Lavender and Tan . . . 70	Maroon and Tan 78
Blue and White 14W	Lavender and White . . . 70W	Copenhagen and Tan . . 55
Brown and Tan 18	Green and Tan 77	Rose and Tan 91
Navy and Tan 23	Green and White . . . 77W	Rose and White 91W

(Page 10)

YORK PLAID BLANKETS—Single

STYLE YORK 2

STYLE YORK 1

STYLE YORK. Single. Size 66 x 80. Ends hemmed.

COLOR COMBINATIONS FOR ALL YORKS

Blue and Tan 14	Lavender and Tan 70	Maroon and Tan 78
Blue and White 14W	Lavender and White . . . 70W	Copenhagen and Tan . . 55
Brown and Tan 18	Green and Tan 77	Rose and Tan 91
Navy and Tan 23	Green and White 77W	Rose and White 91W

(Page 9)

KISMET PLAID BLANKETS—Single

STYLE KISMET 41

STYLE KISMET 52

STYLE KISMET. Single. Size 66 x 80. Ends bound with 3-inch sateen.

COLOR COMBINATIONS FOR ALL KISMETS

Blue and Tan 14	Lavender and Tan 70	Maroon and Tan 78
Blue and White 14W	Lavender and White . . . 70W	Copenhagen and Tan . . 55
Brown and Tan 18	Green and Tan 77	Rose and Tan 91
Navy and Tan 23	Green and White 77W	Rose and White 91W

(Page 11)

SIGNET BLANKETS—Single

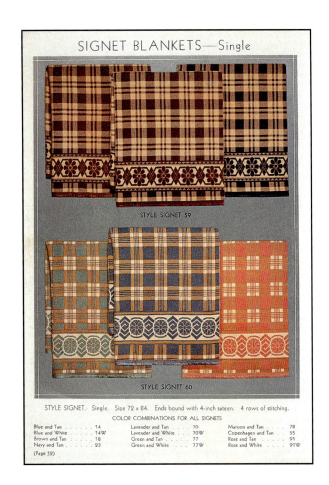

STYLE SIGNET 59

STYLE SIGNET 60

STYLE SIGNET. Single. Size 72 x 84. Ends bound with 4-inch sateen. 4 rows of stitching.

COLOR COMBINATIONS FOR ALL SIGNETS

Blue and Tan 14	Lavender and Tan 70	Maroon and Tan 78
Blue and White 14W	Lavender and White . . . 70W	Copenhagen and Tan . . 55
Brown and Tan 18	Green and Tan 77	Rose and Tan 91
Navy and Tan 23	Green and White 77W	Rose and White 91W

(Page 12)

REVERSIBLE TOP BLANKETS—Single

STYLE PC 1

STYLE PC 1. Single. Size 72 x 84. Ends bound with 3-inch satin. Boxed.

COLORS

Rose and Gold 1	Green and Lavender 2	Blue and Rose 3
Lavender and Gold 4	Green and Rose 5	

(Page 16)

MERLIN BLANKETS—Single

STYLE MERLIN. Single. Size 66 x 80. Ends bound with 3-inch sateen.

COLORS

Rose 1	Blue 2	Green 3
Lavender 4	Peach 5	Rust 6

(Page 13)

CURLEW BLANKETS—Single

STYLE CURLEW 3

STYLE CURLEW 4

STYLE CURLEW. Single. Size 70 x 80. Ends bound with 4-inch sateen. 4 rows of stitching.

COLORS

Rose, Tan Ombre with Dark Red 1	Blue, Tan Ombre with Dark Blue 5
Golden Brown, Tan Ombre with Dark Brown 2	Red, Tan Ombre with Dark Red 6
Lavender, Tan Ombre with Purple 3	Red, Gray Ombre with Dark Red 12
Green, Tan Ombre with Dark Green 4	Blue, Gray Ombre with Dark Blue 13

(Page 18)

108

CURLEW BLANKETS—Single

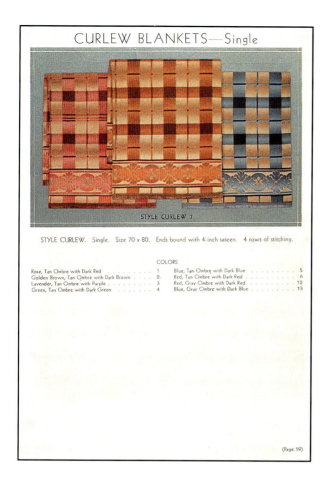

STYLE CURLEW 7

STYLE CURLEW. Single. Size 70 x 80. Ends bound with 4-inch sateen. 4 rows of stitching.

COLORS

Rose, Tan Ombre with Dark Red	1	Blue, Tan Ombre with Dark Blue	5
Golden Brown, Tan Ombre with Dark Brown	2	Red, Tan Ombre with Dark Red	6
Lavender, Tan Ombre with Purple	3	Red, Gray Ombre with Dark Red	12
Green, Tan Ombre with Dark Green	4	Blue, Gray Ombre with Dark Blue	13

HURON BLANKETS—Single

STYLE HURON 1

STYLE HURON 2

STYLE HURON. Single. Size 66 x 80. Ends hemmed.

COLORS

Tan . . . .   Red . . . .   Green . . . .   Navy . . . .

YUKON BLANKETS—Single

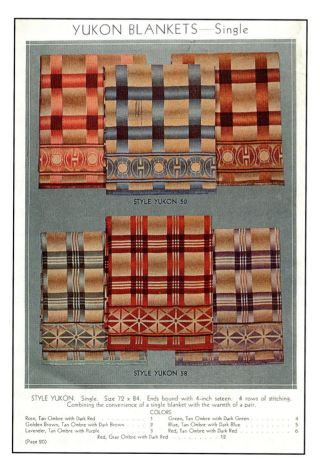

STYLE YUKON 50

STYLE YUKON 58

STYLE YUKON. Single. Size 72 x 84. Ends bound with 4-inch sateen. 4 rows of stitching.
Combining the convenience of a single blanket with the warmth of a pair.

COLORS

Rose, Tan Ombre with Dark Red	1	Green, Tan Ombre with Dark Green	4
Golden Brown, Tan Ombre with Dark Brown	2	Blue, Tan Ombre with Dark Blue	5
Lavender, Tan Ombre with Purple	3	Red, Tan Ombre with Dark Red	6
		Red, Gray Ombre with Dark Red	12

AGAWAM BLANKETS—Single

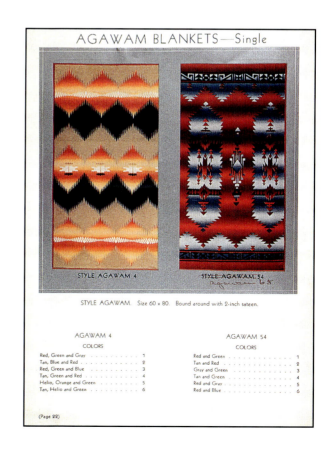

STYLE AGAWAM 4

STYLE AGAWAM 54

STYLE AGAWAM. Size 60 x 80. Bound around with 2-inch sateen.

AGAWAM 4 COLORS		AGAWAM 54 COLORS	
Red, Green and Gray	1	Red and Green	1
Tan, Blue and Red	2	Tan and Red	2
Red, Green and Blue	3	Gray and Green	3
Tan, Green and Red	4	Tan and Green	4
Helio, Orange and Green	5	Red and Gray	5
Tan, Helio and Green	6	Red and Blue	6

AGAWAM BLANKETS—Single

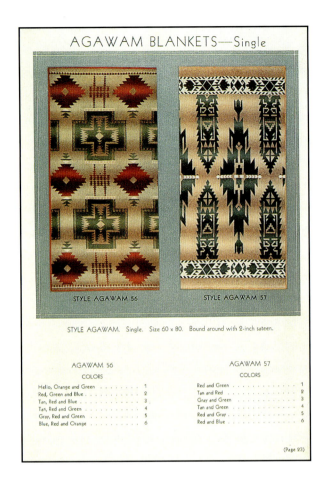

STYLE AGAWAM 56

STYLE AGAWAM 57

STYLE AGAWAM. Single. Size 60 x 80. Bound around with 2-inch sateen.

AGAWAM 56 COLORS		AGAWAM 57 COLORS	
Helio, Orange and Green	1	Red and Green	1
Red, Green and Blue	2	Tan and Red	2
Tan, Red and Blue	3	Gray and Green	3
Tan, Red and Green	4	Tan and Green	4
Gray, Red and Green	5	Red and Gray	5
Blue, Red and Orange	6	Red and Blue	6

(Page 23)

INCA BLANKETS—Single

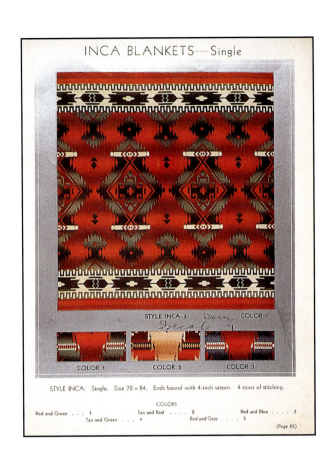

STYLE INCA 3   COLOR 1

COLOR 1   COLOR 2   COLOR 3

STYLE INCA. Single. Size 72 x 84. Ends bound with 4-inch sateen. 4 rows of stitching.

COLORS

Red and Green . . . 1   Tan and Red 2   Red and Blue 3
Tan and Green . . . 4   Red and Gray 5

(Page 25)

WIGWAM BLANKETS—Single

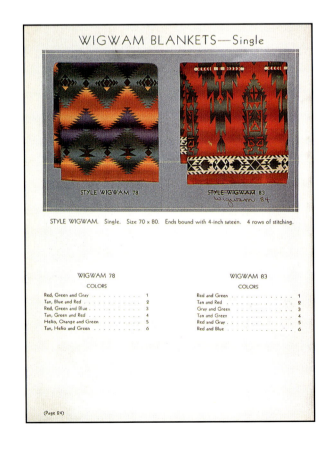

STYLE WIGWAM 78

STYLE WIGWAM 83

STYLE WIGWAM. Single. Size 70 x 80. Ends bound with 4-inch sateen. 4 rows of stitching.

WIGWAM 78 COLORS		WIGWAM 83 COLORS	
Red, Green and Gray	1	Red and Green	1
Tan, Blue and Red	2	Tan and Red	2
Red, Green and Blue	3	Gray and Green	3
Tan, Green and Red	4	Tan and Green	4
Helio, Orange and Green	5	Red and Gray	5
Tan, Helio and Green	6	Red and Blue	6

(Page 24)

INCA BLANKETS—Single

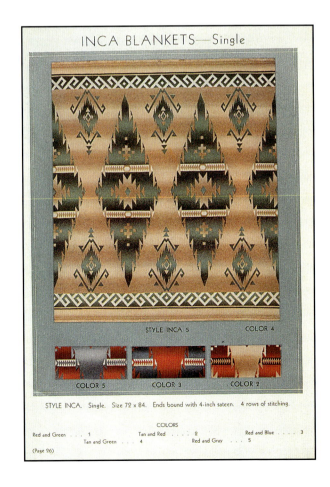

STYLE INCA 5   COLOR 4

COLOR 5   COLOR 3   COLOR 2

STYLE INCA. Single. Size 72 x 84. Ends bound with 4-inch sateen. 4 rows of stitching.

COLORS

Red and Green . . . 1   Tan and Red 2   Red and Blue 3
Tan and Green . . . 4   Red and Gray 5

(Page 26)

110

The color combination on this Indian design Ombre was used extensively in the 1930s, patternization side shown. Courtesy of Bearwallow Mountain Traders, Judy Hudson & Norwood Barnes.

Colorization side of the same blanket. Courtesy of Bearwallow Mountain Traders, Judy Hudson & Norwood Barnes.

Brilliant execution of color and design on this Ombre Beacon, mid-1930s. Courtesy of R. Greg Otto, Portland, Oregon.

An exceptional Ombre design combination, mid-1930s. Courtesy of R. Greg Otto, Portland, Oregon.

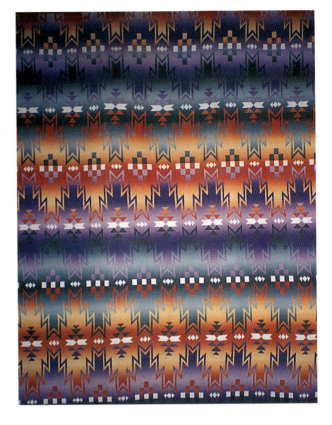

An Ombre blanket ready for a romantic evening by the fireplace, "c. 1930." Author's collection.

The design and colors of this Ombre create a third dimensional look, "c. 1930s." Courtesy of Bearwallow Mountain Traders, Judy Hudson & Norwood Barnes.

Classic Beacon tan, red and green combination gives a distinctive look to this Ombre shawl, mid to late 1930s. Courtesy of Bearwallow Mountain Traders, Judy Hudson & Norwood Barnes.

A mid to late-1930s Ombre design with an autumn color combination. Courtesy of Laura Fisher/Antique Quilts and Americana.

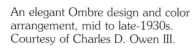

An elegant Ombre design and color arrangement, mid to late-1930s. Courtesy of Charles D. Owen III.

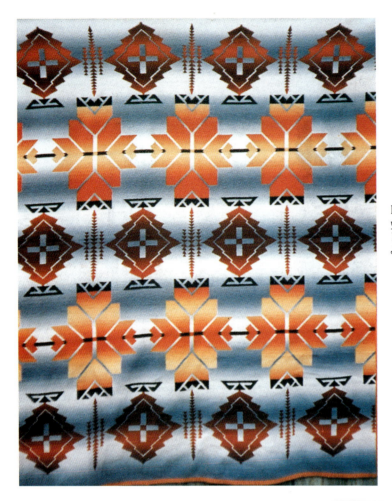

Indian Ombre motif made from Beacon's own dyed yarns in a very attractive color arrangement, late-1930s. Courtesy of Bearwallow Mountain Traders, Judy Hudson & Norwood Barnes.

Red, gray and shades of orange and black in the same Ombre pattern give an entirely different look, late 1930s. Courtesy of Bearwallow Mountain Traders, Judy Hudson & Norwood Barnes.

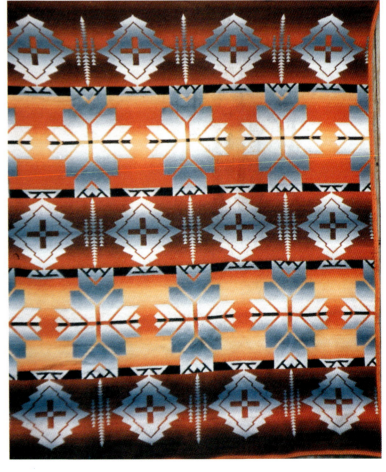

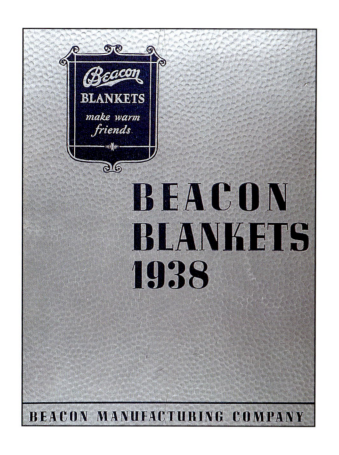

BEACON BLANKETS 1938

BEACON MANUFACTURING COMPANY

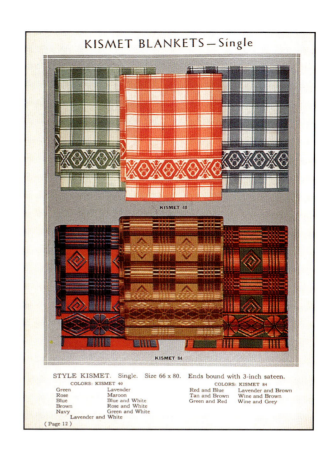

KISMET BLANKETS — Single

KISMET 40

KISMET 84

STYLE KISMET. Single. Size 66 x 80. Ends bound with 3-inch sateen.

COLORS: KISMET 40		COLORS: KISMET 84	
Green	Lavender	Red and Blue	Lavender and Brown
Rose	Maroon	Tan and Brown	Wine and Brown
Blue	Blue and White	Green and Red	Wine and Grey
Brown	Rose and White		
Navy	Green and White		
	Lavender and White		

(Page 12)

YORK BLANKETS — Single

YORK 7

YORK 8

STYLE YORK. Single. Size 66 x 80. Ends hemmed.

COLORS: YORK 7		COLORS: YORK 8	
Green and Red	Red and Blue	Brown	Navy
Tan and Brown	Lavender and Brown	Red	Maroon
Wine and Brown	Wine and Grey		Green

(Page 11)

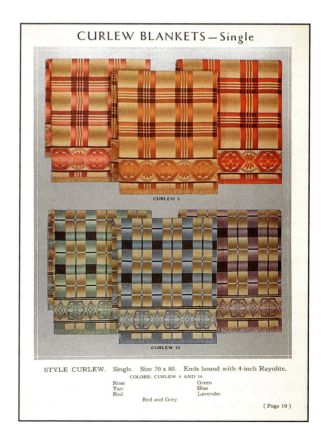

CURLEW BLANKETS — Single

CURLEW 4

CURLEW 16

STYLE CURLEW. Single. Size 70 x 80. Ends bound with 4-inch Rayolite.

COLORS: CURLEW 4 AND 16		
Rose		Green
Tan		Blue
Red		Lavender
	Red and Grey	

(Page 19)

YUKON BLANKETS — Single

YUKON 60

YUKON 64

STYLE YUKON. Single. Size 72 x 84. Ends bound with 4-inch sateen.
COLORS: YUKON 60 AND 64

Blue   Rose
Green   Lavender
Tan   Red
        Red and Grey

AGAWAM BLANKETS — Single

STYLE AGAWAM 4   STYLE AGAWAM 57

STYLE AGAWAM. Single. Size 60 x 80. Bound around with 2-inch sateen.
COLORS: AGAWAM 4   COLORS: AGAWAM 57
Tan   Red
Red   Tan and Red
Blue   Blue
Lavender   Tan and Green

HURON BLANKETS — Single

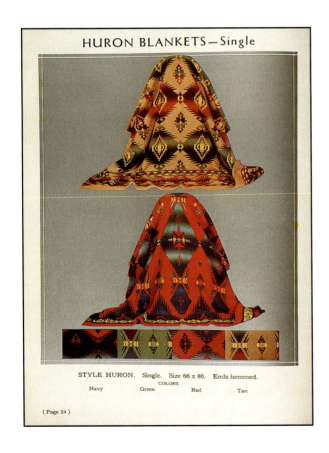

STYLE HURON. Single. Size 66 x 80. Ends hemmed.
COLORS
Navy   Green   Red   Tan

AGAWAM BLANKETS — Single

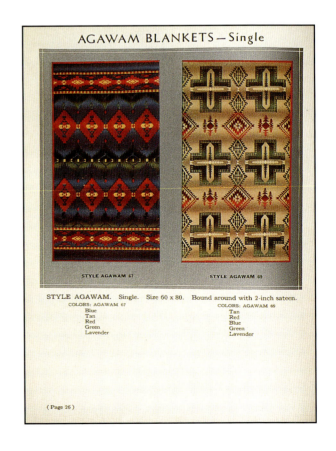

STYLE AGAWAM 67   STYLE AGAWAM 69

STYLE AGAWAM. Single. Size 60 x 80. Bound around with 2-inch sateen.
COLORS: AGAWAM 67   COLORS: AGAWAM 69
Blue   Tan
Tan   Red
Red   Blue
Green   Green
Lavender   Lavender

116

STYLE INCA 10

STYLE INCA. Single. Size 72 x 84. Ends bound with 4-inch sateen.

COLORS

Red
Blue

Green
Lavender

Tan

(Page 29)

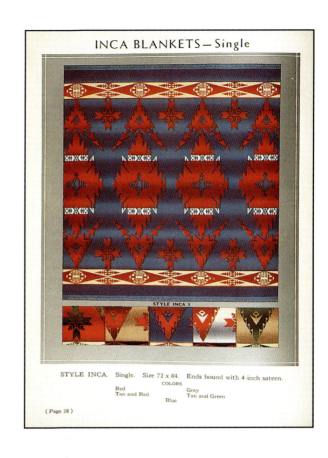

STYLE INCA 9

STYLE INCA. Single. Size 72 x 84. Ends bound with 4-inch sateen.

COLORS

Red
Tan and Red

Grey
Tan and Green

Blue

(Page 28)

WIGWAM BLANKETS—Single

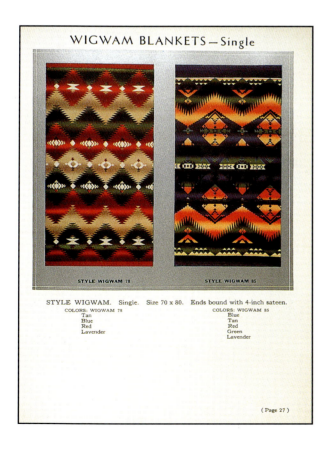

STYLE WIGWAM 78   STYLE WIGWAM 85

STYLE WIGWAM. Single. Size 70 x 80. Ends bound with 4-inch sateen.

COLORS: WIGWAM 78
Tan
Blue
Red
Lavender

COLORS: WIGWAM 85
Blue
Tan
Red
Green
Lavender

(Page 27)

Beacon BLANKETS make warm friends

BEACON BLANKETS 1939

BEACON MANUFACTURING COMPANY

117

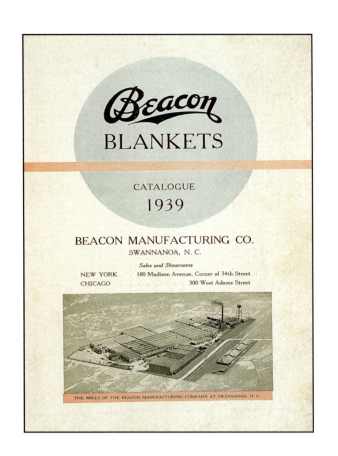

Beacon
BLANKETS

CATALOGUE
1939

BEACON MANUFACTURING CO.
SWANNANOA, N. C.

Sales and Showrooms

NEW YORK . . . . . 180 Madison Avenue, Corner of 34th Street
CHICAGO . . . . 300 West Adams Street

THE MILLS OF THE BEACON MANUFACTURING COMPANY AT SWANNANOA, N. C.

P—PLAID BLANKETS—Pairs

P 176

P 161

STYLE P. Pairs. Size 72 x 84. Ends bound with four-inch sateen.

COLORS: P 176		COLORS: P 161	
Blue	Green	Tan and Brown	Lavender and Brown
Brown	Navy	Green and Red	Blue and Red
Lavender	Rose	Rust and Green	
Maroon	Wine		

(Page 12)

P—PLAID BLANKETS—Pairs

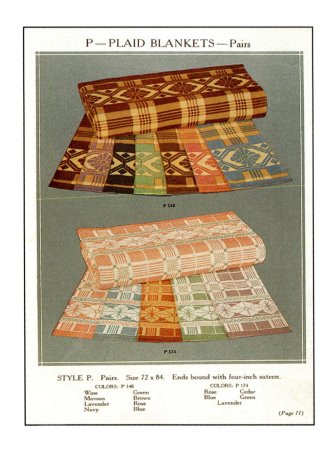

P 148

P 174

STYLE P. Pairs. Size 72 x 84. Ends bound with four-inch sateen.

COLORS: P 148		COLORS: P 174	
Wine	Green	Rose	Cedar
Maroon	Brown	Blue	Green
Lavender	Rose	Lavender	
Navy	Blue		

(Page 11)

MIDWAY BLANKETS—Singles

MIDWAY 37

MIDWAY 38

STYLE MIDWAY. Single. Size 60 x 80. Ends hemmed.

COLORS: MIDWAY 37		COLORS: MIDWAY 38	
Red and Blue	Blue and Red	Tan and Brown	Green and Dark Green
Rust and Green	Tan and Brown	Rose and Maroon	Wine and Green
Green and Red	Wine and Grey	Blue and Red	Black and Red
		Lavender and Dark Lavender	

(Page 13)

YORK BLANKETS — Singles

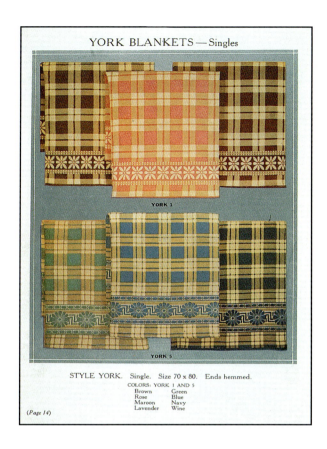

YORK 1

YORK 5

STYLE YORK. Single. Size 70 x 80. Ends hemmed.

COLORS: YORK 1 AND 5

Brown	Green
Rose	Blue
Maroon	Navy
Lavender	Wine

KISMET BLANKETS — Singles

KISMET 40

KISMET 84

STYLE KISMET. Single. Size 70 x 80. Ends bound with three-inch sateen.

COLORS: KISMET 40		COLORS: KISMET 84	
Green	Navy	Red and Blue	Blue and Red
Rose	Wine	Tan and Brown	Rust and Green
Blue	Lavender	Green and Red	Wine and Grey
Brown	Maroon		

YORK BLANKETS — Singles

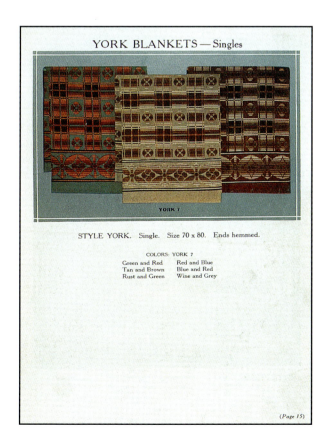

YORK 7

STYLE YORK. Single. Size 70 x 80. Ends hemmed.

COLORS: YORK 7

Green and Red	Red and Blue
Tan and Brown	Blue and Red
Rust and Green	Wine and Grey

KISMET BLANKETS — Singles

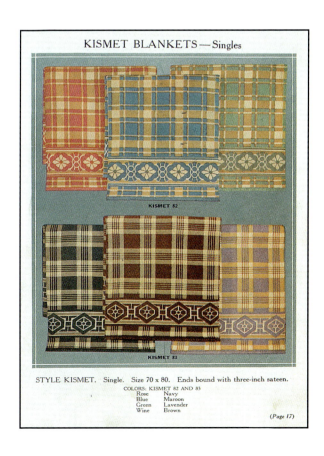

KISMET 82

KISMET 83

STYLE KISMET. Single. Size 70 x 80. Ends bound with three-inch sateen.

COLORS: KISMET 82 AND 83

Rose	Navy
Blue	Maroon
Green	Lavender
Wine	Brown

MISTRAL BLANKETS — Singles

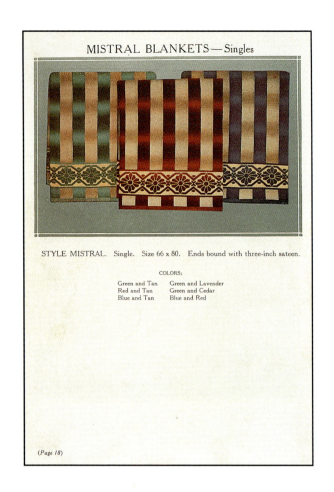

STYLE MISTRAL.   Single.   Size 66 x 80.   Ends bound with three-inch sateen.

COLORS:

Green and Tan	Green and Lavender
Red and Tan	Green and Cedar
Blue and Tan	Blue and Red

YUKON BLANKETS — Singles

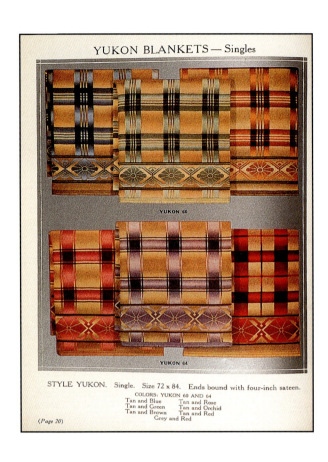

YUKON 60

YUKON 64

STYLE YUKON.   Single.   Size 72 x 84.   Ends bound with four-inch sateen.

COLORS: YUKON 60 AND 64

Tan and Blue	Tan and Rose
Tan and Green	Tan and Orchid
Tan and Brown	Tan and Red
Grey and Red	

CURLEW BLANKETS — Singles

CURLEW 4

CURLEW 16

STYLE CURLEW.   Single.   Size 72 x 84.   Ends bound with four-inch rayon taffeta.

COLORS: CURLEW 4 AND 16

Tan and Rose	Tan and Green
Tan and Brown	Tan and Blue
Tan and Red	Tan and Orchid
Grey and Red	

PC 9 BLANKETS — Singles

STYLE PC 9.   Single.   Size 72 x 84.
Bound around with four-inch rayon taffeta.   Boxed.

COLORS:

Blue	Peach
Green	Rose
Cedar	Orchid

HURON BLANKETS — Singles

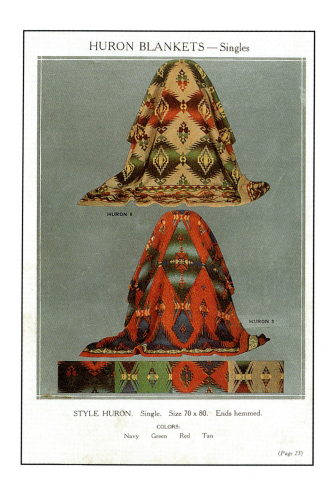

STYLE HURON. Single. Size 70 x 80. Ends hemmed.

COLORS:

Navy   Green   Red   Tan

WIGWAM BLANKETS — Singles

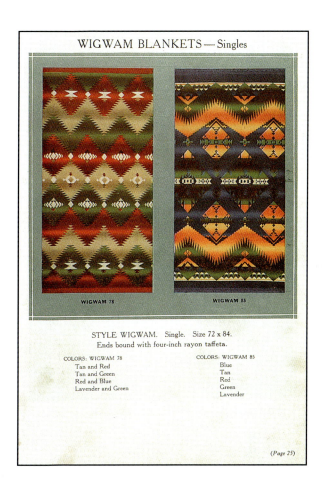

STYLE WIGWAM. Single. Size 72 x 84.
Ends bound with four-inch rayon taffeta.

COLORS: WIGWAM 78	COLORS: WIGWAM 85
Tan and Red	Blue
Tan and Green	Tan
Red and Blue	Red
Lavender and Green	Green
	Lavender

AGAWAM BLANKETS — Singles

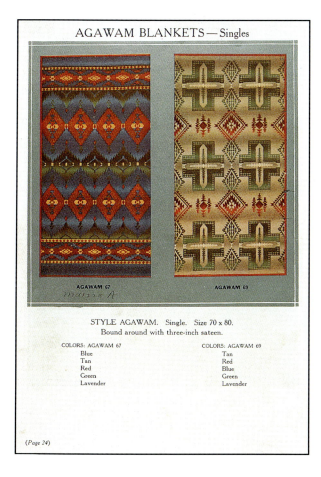

STYLE AGAWAM. Single. Size 70 x 80.
Bound around with three-inch sateen.

COLORS: AGAWAM 67	COLORS: AGAWAM 69
Blue	Tan
Tan	Red
Red	Blue
Green	Green
Lavender	Lavender

INCA BLANKETS — Singles

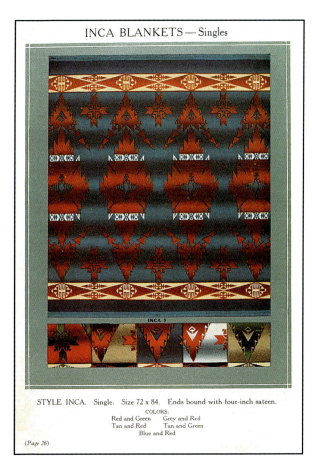

STYLE INCA. Single. Size 72 x 84. Ends bound with four-inch sateen.

COLORS:

Red and Green   Grey and Red
Tan and Red   Tan and Green
Blue and Red

INCA BLANKETS — Singles

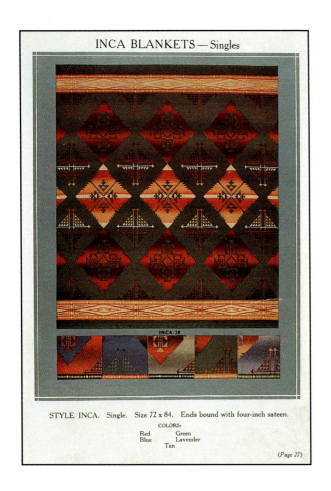

STYLE INCA. Single. Size 72 x 84. Ends bound with four-inch sateen.

COLORS:

Red	Green
Blue	Lavender
Tan	

(Page 27)

MERLIN BLANKETS — Singles
25% WOOL

STYLE MERLIN. Single. Size 70 x 80. Ends bound with three-inch rayon taffeta.

COLORS:

Green and Rose	Rose and Green
Lavender and Green	Blue and Rose
Wine and Red	Cedar and Tan
Royal Blue and Light Blue	Moss Green and Light Green

(Page 29)

Beacon
25% WOOL
BLANKETS

(See pages 29 to 34)

(Page 28)

NOMAD BLANKETS — Singles
25% WOOL

NOMAD 18

NOMAD 19

STYLE NOMAD. Single. Size 72 x 84. Ends bound with four-inch rayon taffeta.

COLORS: NOMAD 18 AND 19

Blue	Lavender
Rose	Navy
Wine	Green
Brown	

(Page 30)

ROMA BLANKETS — Singles
25% WOOL

STYLE ROMA. Single. Size 72 x 84. Ends bound with four-inch rayon taffeta.

COLORS:
Royal Blue   Wine
Cedar   Moss Green
Lavender

(Page 31)

CASCO BLANKETS — Singles
25% WOOL

CASCO 45

STYLE CASCO 45. Single. Size 70 x 80. Ends bound with four-inch rayon taffeta.

COLORS:
Blue   Green   Tan   Red

(Page 33)

LOTUS BLANKETS — Singles
25% WOOL

STYLE LOTUS. Single. Size 72 x 84. Ends bound with four-inch rayon taffeta.

COLORS:
Green and Rose   Blue and Rose
Rose and Green   Lavender and Green
Moss Green and Maize   Cedar and Maize
Wine and Maize   Royal Blue and Maize

(Page 32)

CASCO BLANKETS — Singles
25% WOOL

CASCO 46

STYLE CASCO 46. Single. Size 70 x 80.
Ends bound with four-inch rayon taffeta.

COLORS:
Green   Blue   Tan   Red

(Page 34)

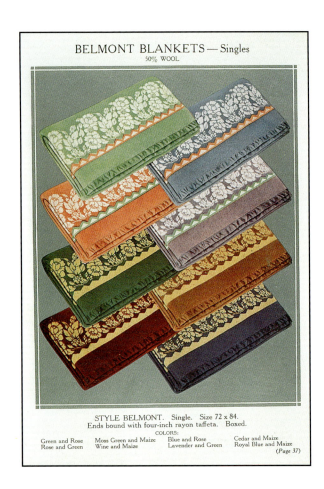

BELMONT BLANKETS — Singles
50% WOOL

STYLE BELMONT.   Single.   Size 72 x 84.
Ends bound with four-inch rayon taffeta.   Boxed.

COLORS:

Green and Rose	Moss Green and Maize	Blue and Rose	Cedar and Maize
Rose and Green	Wine and Maize	Lavender and Green	Royal Blue and Maize

(Page 37)

Inca design Ombre used often over the years, late-1930s. Author's collection. $75-125.

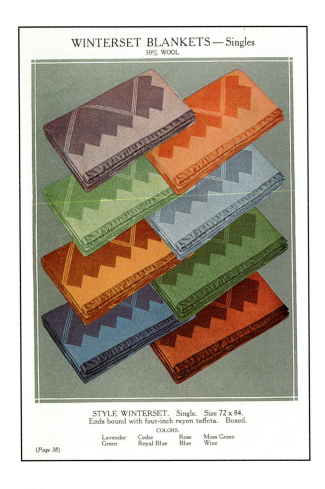

WINTERSET BLANKETS — Singles
50% WOOL

STYLE WINTERSET.   Single.   Size 72 x 84.
Ends bound with four-inch rayon taffeta.   Boxed.

COLORS:

Lavender	Cedar	Rose	Moss Green
Green	Royal Blue	Blue	Wine

(Page 38)

Another of Beacon's endless color combinations, Ombre with a strong graphic design, late-1930s. Courtesy of Bearwallow Mountain Traders, Judy Hudson & Norwood Barnes.

An exciting and vivid Ombre shawl, late-1930s. Courtesy of Bearwallow Mountain Traders, Judy Hudson & Norwood Barnes.

Reverse side of previous vivid Ombre shawl. Courtesy of Bearwallow Mountain Traders, Judy Hudson & Norwood Barnes.

Kismet blanket design came in eleven different color combinations, 1935/1936. Author's collection. $40-75.

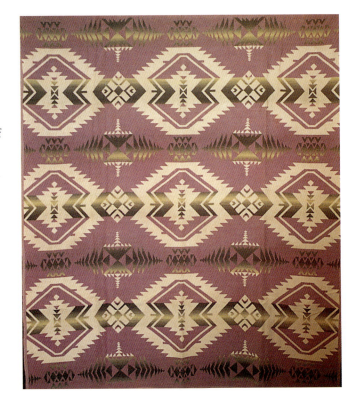

Indian Ombre design in a very "forward" color combination for its time, late-1930s. Courtesy of Laura Fisher/Antique Quilts and Americana.

The bottom of this Ombre Indian design blanket was not lost in a campfire! mid to late-1930s. Courtesy of Laura Fisher/Antique Quilts and Americana.

Toba Indian design economy blanket, late-1930s. Author's collection. $125-195.

Strips of Beacon blanket fabric sewn together to form a blanket. Thought to be a salesman's sample. Notice several Ombre pieces. Late 30's to early 40's. Courtesy of Sam and Denise Kennedy of Cisco's, Coeur d'Alene, Idaho.

Montgomery Ward Catalog, Spring and Summer 1948. "B" Ward's best novelty blanket, plaid pattern 72x84, weight 2.75 lbs., cotton colors: rust/green, wine/tan, navy/tan, each $4.39. "D" Indian style blanket, 66x80 weight 2.25 lbs., cotton colors: red, green or navy predominating, each $3.39. Courtesy of Montgomery Ward Collection, American Heritage Center, University of Wyoming.

Transition Period: 1940-1957

The 1941 through 1948 catalogs are absent, presumably because during the Second World War Beacon was producing blankets for the military. Modern manufacturing equipment was installed starting around 1950, and by 1954/55 the cotton blankets had been completely replaced by rayon blends. The *Ombre* design process was no longer feasible.

Wardrobe closet brim full of beautiful Beacon blankets at Cisco. Courtesy of Sam and Denise Kennedy of Cisco's, Coeur d'Alene, Idaho.

A mixture of Indian designs combine to make an attractive blanket, not an Ombre, late-1940s. Author's collection. $75-150.

Same blanket design but in different colors, late-1940s. Courtesy of Sam and Denise Kennedy of Cisco's, Coeur d'Alene, Idaho.

Beacon's ability to make a beautiful product is apparent in this fine blanket, late-1940s. Courtesy of R. Greg Otto, Portland, Oregon.

This Indian design was originally produced in the 1930s as an Ombre but this particular blanket is not an Ombre, late-1940s. Courtesy of Laura Fisher/Antique Quilts and Americana.

The Indian design on this blanket is not an Ombre, but its design placement makes it appear that it is, late-1940s. Courtesy of Gary Holt/Steve Christianson, Gary Holt collection.

Multiple Indian design elements and strong colors make an exceptional blanket, late-1940s. Courtesy of Laura Fisher/Antique Quilts and Americana.

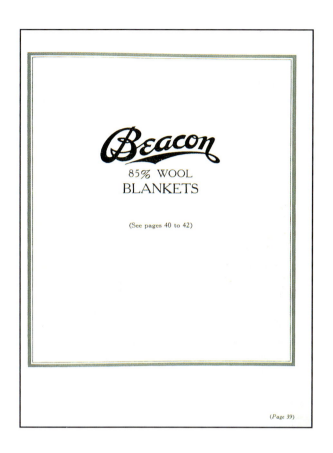

(See pages 40 to 42)

Beacon
85% WOOL
BLANKETS

Beacon
BLANKETS
1940
BEACON MANUFACTURING COMPANY

SACHEM BLANKETS — Singles
85% WOOL

STYLE SACHEM. Single. Size 72 x 84.
Ends bound with four-inch acetate satin. Boxed.
COLORS:
Red   Blue   Tan   Green

FAUNA PLAID BLANKETS · Pairs

FAUNA 3

STYLE FAUNA. Pairs. Size 70x90. Ends bound with four-inch sateen.
COLORS: FAUNA 3

Red and Blue   Tan and Brown
Royal and Red   Green and Red
            Green and Rust
            Wine and Gray

MIDWAY BLANKETS · Singles

MIDWAY 37

MIDWAY 94

STYLE MIDWAY.   Single.   Size 60x80.   Ends hemmed.

COLORS: MIDWAY 37

Red and Blue	Wine and Gray
Green and Red	Royal and Red
Tan and Brown	Rust and Green

COLORS: MIDWAY 94

Green and Red	Brown and Green
Navy and Red	Black and Green
Lavender and Green	Black and Wine

MIDWAY 95.   Not Illustrated.   COLORS: Navy, Rust, Maroon, Green, Rose, Blue.

Page 12

FAUNA PLAID BLANKETS · Pairs

FAUNA 1

FAUNA 2

STYLE FAUNA.   Pairs.   Size 70x80.   Ends bound with four-inch sateen.

COLORS: FAUNA 1 & 2

Blue	Lavender	Rust	Navy
Rose	Green	Wine	Maroon
	Brown		

Page 8

CURLEW BLANKETS · Singles

CURLEW 4

CURLEW 19

STYLE CURLEW.   Single.   Size 72x84.   Ends bound with four-inch rayon taffeta.

COLORS: CURLEW 4 AND 19

Tan and Rose	Tan and Green
Tan and Brown	Tan and Blue
Tan and Red	Tan and Orchid
Gray and Red	
Green and Cedar	

Page 18

HURON BLANKETS · Singles

HURON 8

HURON 10

STYLE HURON.   Single.   Size 70x80.   Ends hemmed.

COLORS:

Navy	Green	Red	Tan

Page 23

132

TOBA BLANKETS · Singles

TOBA 35

TOBA 41

STYLE TOBA. Single. Size 60x80. Ends hemmed.

COLORS:

Green   Red   Tan   Navy   Gray

Page 22

YORK 26   YORK 112   YORK 30

NOVELTY JACQUARD PLAID SINGLE

STYLE YORK 26   70" x 80"   Hemmed ends
    Packing: 30 to carton assorted — 4 blue-tan, 4 rose-tan, 8 navy-red, 4 wine-tan, 4 rust-tan, 4 green-peach

STYLE YORK 30   70" x 80"   Hemmed ends
    Packing: 30 to carton assorted — 4 tan-blue, 4 red-blue, 6 red-gray, 4 green-peach, 6 navy-red, 6 red-tan

STYLE YORK 112   70" x 80"   Hemmed ends
    Packing: 30 to carton assorted — 8 lt. blue-dark blue, 8 lt. rose-dark rose, 6 lt. green-dark green, 8 lt. gray-cherry

(8)

Beacon

BLANKETS
"make warm friends"

1950

BEACON MANUFACTURING COMPANY

KISMET 112   KISMET 26   KISMET 30

NOVELTY JACQUARD PLAID SINGLE

STYLE KISMET 26   70" x 80"   Bound 3" acetate satin
    Packing: 30 to carton assorted — 4 blue-tan, 4 rose-tan, 8 navy-red, 6 wine-tan, 4 rust-tan, 4 green-peach

STYLE KISMET 30   70" x 80"   Bound 3" acetate satin
    Packing: 30 to carton assorted — 4 tan-blue, 4 red-blue, 6 red-gray, 4 green-peach, 6 navy-red, 6 red-tan

STYLE KISMET 112   70" x 80"   Bound 3" acetate satin
    Packing: 30 to carton assorted — 8 lt. blue-dark blue, 8 lt. rose-dark rose, 6 lt. green-dark green, 8 lt. gray-cherry

(9)

SOUTHFIELD 148   YUKON 101   INCA 10

NOVELTY JACQUARD PLAID SINGLE

STYLE SOUTHFIELD 148   72" x 84"   Bound 4" acetate satin
  Packing: 30 to carton — 4 rose, 4 blue, 4 rust, 2 navy,
  6 wine, 6 navy-red, 4 green-peach
STYLE YUKON 101   72" x 84"   Bound 4" acetate satin
  Packing: 30 to carton assorted — 6 tan-red, 8 tan-rose,
  6 tan-blue, 6 gray-red, 4 green-rust

NOVELTY JACQUARD INDIAN DESIGN SINGLE

STYLE INCA 10   72" x 84"   Bound 4" acetate satin
  Packing: 30 to carton — 6 green, 8 blue, 8 red, 8 tan

(10)

HOPI   SACHEM 107   SACHEM 102

INDIAN DESIGN BLANKET SINGLE

STYLE SACHEM 102 (Upper)   64" x 76"   Hemmed ends
STYLE HOPI (Center)   60" x 76"   Hemmed ends
STYLE SACHEM 107 (Lower)   64" x 76"   Hemmed ends
  Packing: 30 one pattern to carton, assorted — 8 blue,
  8 green, 8 red, 6 tan

(12)

5% WOOL, 65% RAYON, AND 30% COTTON
SOLID COLOR JACQUARD BORDER SINGLE

STYLE MERLIN 22   70" x 80"   Bound 3" acetate satin
  Packing: 30 to carton — 4 blue, 4 rose, 2 green, 2 maize, 4 hunter
  green, 4 geranium red, 2 gray with cherry, 4 wine, 4 royal

JACQUARD COTTON SINGLE

STYLE VINES No. 2   70" x 80"   Bound 3" acetate satin
  Packing: 30 to carton — 8 geranium red-gray, 6 hunter green-
  gray, 6 royal-red, 4 cedar-peach, 6 maroon-gray

(11)

TOBA 52   TOBA 75

INDIAN DESIGN BLANKET SINGLE

STYLE TOBA 52   70" x 80"   60 x 80   Hemmed ends
STYLE TOBA 75   70" x 80"   60 x 80   Hemmed ends
  Packing: 30 of a style to carton, assorted — 6 red, 6 navy,
  6 dark green, 6 tan, 6 maroon

(13)

134

HURON 72   HURON 73

INDIAN DESIGN BLANKET SINGLE

STYLE HURON 72   70" x 80"   Hemmed ends
STYLE HURON 73   70" x 80"   Hemmed ends
Packing: 30 of one style to carton, assorted — 6 red,
6 gray, 6 green, 6 tan, 6 navy

(14)

AGAWAM 72   AGAWAM 73

INDIAN DESIGN BLANKET SINGLE

STYLE AGAWAM 72   70" x 80"   Bound 3" acetate satin
STYLE AGAWAM 73   70" x 80"   Bound 3" acetate satin
Packing: 30 of one style to carton, assorted — 6 red,
6 gray, 6 green, 6 tan, 6 navy

(15)

TEMPLE   DEN

10% WOOL, 65% RAYON, AND 25% COTTON SINGLE

STYLE DEN   72" x 84"   Overlocked stitched ends, Boxed
Packing: 12 or 24 to carton — white ground striped color border

STYLE TEMPLE   72" x 90"   Bound 5" acetate satin, Boxed
Packing: 12 or 24 to carton — 6 wine, 6 green, 6 gray, 6 royal

(17)

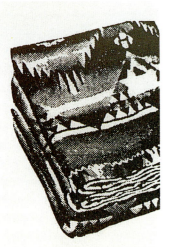

For Cool Summe

INDIA
and PLA
BLANKE
2.73

They're so good-look
inexpensive you'll wa
than one! Soft and
single cotton blanket
inch, in gay Indian a
designs! Swell NOW f

Lebanon Express newspaper, Lebanon, Oregon, November 28, 1952. J.C. Penny ads: "Indian and Plaid Blankets soft and sturdy, single cotton blankets 66x80, swell NOW for camping, $2.73". Courtesy J. C. Penny Company, Plano, Texas.

An excellent color mix with white detail accentuating the design, early-1950's. Courtesy of Sam and Denise Kennedy of Cisco's, Coeur d'Alene, Idaho.

Indian Design Cotton Blanket, 70x80, "Bright and cheery as a camp-fire, $2.49" Courtesy J. C. Penny Com-pany, Plano, Texas.

Indian Design ...

COTTOI
BLANKE
2.49

Bright and cheery as a c
fire! A 70x80 inch single
ton blanket in handsome

With matching blanket. Author's collection. $65-175.

136

Indian design orange-fringed shawl, not an Ombre, early-1950s. Courtesy of John & Cindy Speare.

Indian designs in three colors without the Ombre process, mid-1950s. Author's collection. $95-150.

An exceptionally beautiful blanket, but the design is not Ombre, late-1940's to early-1950's. Courtesy of Bearwallow Mountain Traders, Judy Hudson & Norwood Barnes.

A four color Indian design blanket without the Ombre process, "c. 1950's." Courtesy of Bearwallow Mountain Traders, Judy Hudson & Norwood Barnes.

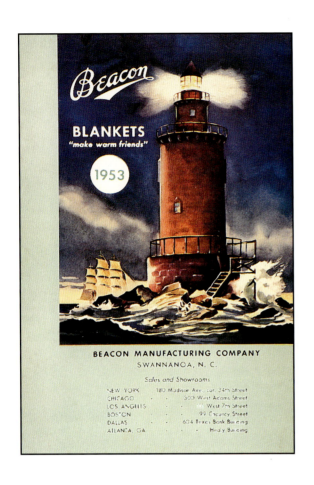

Beacon

BLANKETS
"make warm friends"

1953

BEACON MANUFACTURING COMPANY
SWANNANOA, N. C.

Sales and Showrooms

NEW YORK	180 Madison Ave., cor. 34th Street
CHICAGO	300 West Adams Street
LOS ANGELES	West 7th Street
BOSTON	99 Chauncy Street
DALLAS	604 Texas Bank Building
ATLANTA, GA	Healy Building

MOHAWK

SACHEM 107

INDIAN DESIGN BLANKET SINGLE

STYLE MOHAWK	54" x 72"	Hemmed Ends
STYLE SACHEM 107	64" x 76"	Hemmed Ends

Packed 30 One Style to Carton, Unbagged
Assorted, 8 Blue, 8 Green, 8 Red, 6 Tan

(11)

SOUTHFIELD 148   YUKON 101   INCA 10

NOVELTY JACQUARD PLAID SINGLE

STYLE SOUTHFIELD 148 72" x 84"   Bound 4" Acetate Satin
Packed 30 to Carton, Bagged
6 each Blue, Rose, Rust, Wine, Green and Peach

STYLE YUKON 101   72" x 84"   Bound 4" Acetate Satin
Packed 30 to Carton, Bagged
10 Tan and Rose, 8 Tan and Blue, 7 Gray and Red, 5 Green and Rust

NOVELTY JACQUARD INDIAN DESIGN SINGLE

STYLE INCA 10   72" x 84"   Bound 4" Acetate Satin
Packed 30 to Carton, Bagged
8 Tan, 8 Red, 8 Blue, 6 Green

(9)

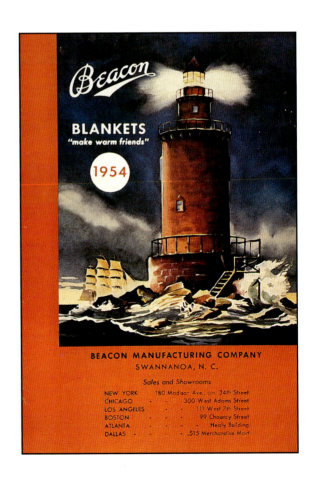

Beacon

BLANKETS
"make warm friends"

1954

BEACON MANUFACTURING COMPANY
SWANNANOA, N. C.

Sales and Showrooms

NEW YORK	180 Madison Ave., cor. 34th Street
CHICAGO	300 West Adams Street
LOS ANGELES	111 West 7th Street
BOSTON	99 Chauncy Street
ATLANTA	Healy Building
DALLAS	515 Merchandise Mart

138

MOHAWK 2

SACHEM 107

INDIAN DESIGN BLANKET SINGLE

STYLE MOHAWK 2   54" x 72"   Hemmed Ends
Packed 30 to Carton, Unbagged
Assorted, 6 Light Blue, 8 Red, 8 Dark Green, 8 Wine

STYLE SACHEM 107   64" x 76"   Hemmed Ends
Packed 30 to Carton, Unbagged
Assorted, 6 Light Blue, 8 Dark Green, 8 Red, 8 Tan

(11)

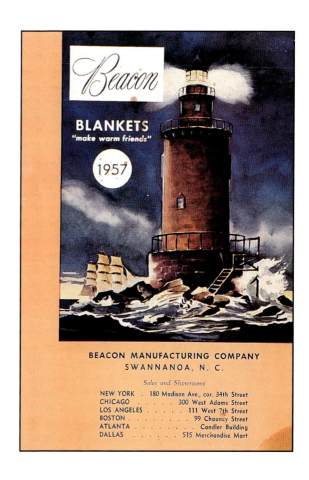

Beacon

BLANKETS
"make warm friends"

1957

BEACON MANUFACTURING COMPANY
SWANNANOA, N. C.

Sales and Showrooms

NEW YORK . . . 180 Madison Ave., cor. 34th Street
CHICAGO . . . . . . 300 West Adams Street
LOS ANGELES . . . . . 111 West 7th Street
BOSTON . . . . 99 Chauncy Street
ATLANTA . . . . Candler Building
DALLAS . . . . . . 515 Merchandise Mart

CHIEF

HURON 72

INDIAN DESIGN BLANKET SINGLE

STYLE CHIEF   70" x 80"   Hemmed Ends
STYLE WIGWAM   70" x 80"   Bound 4" Acetate Satin
Packed 30 One Style to Carton, Bagged
Assorted, 8 Blue, 8 Green, 6 Gray, 8 Red

STYLE HURON 72   70" x 80"   Hemmed Ends
STYLE AGAWAM 72   70" x 80"   Bound 4" Acetate Satin
Packed 30 One Style to Carton, Bagged
Assorted, 8 Blue, 8 Green, 8 Red, 6 Tan

(13)

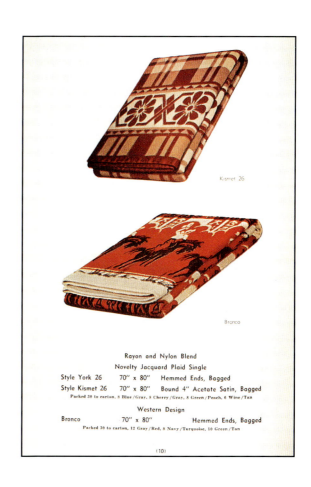

Kismet 26

Bronco

Rayon and Nylon Blend
Novelty Jacquard Plaid Single

Style York 26   70" x 80"   Hemmed Ends, Bagged
Style Kismet 26   70" x 80"   Bound 4" Acetate Satin, Bagged
Packed 30 to carton, 8 Blue /Gray, 8 Cherry /Gray, 8 Green /Peach, 6 Wine /Tan

Western Design

Bronco   70" x 80"   Hemmed Ends, Bagged
Packed 30 to carton, 12 Gray /Red, 8 Navy /Turquoise, 10 Green /Tan

(10)

Mohawk 2

Sachem 139

Rayon and Nylon Blend
Indian Design Blanket Single

Style Mohawk 2   54" x 72"   Hemmed Ends, Bulk
Packed 30 to carton, 6 Blue, 8 Hunter Green, 8 Red, 8 Wine

Style Sachem 139   64" x 76"   Hemmed Ends, Bulk
Packed 30 to carton, 8 Gray, 8 Red, 7 Blue, 7 Green

(11)

LEOPARD

TIGER

NOVELTY JACQUARD ROBE SINGLE

STYLE LEOPARD   54" x 72"   Hemmed Ends
STYLE TIGER   54" x 72"   Hemmed Ends
Color Brown only
Packed 30 to Carton, Unbagged

(10)

The leopard design blanket pictured here appeared in Beacon catalogs for almost ten years. No one remembers why. Was it because it was a top seller or was it because it didn't sell quickly? The leopard design holds the record for the longest running blanket in Beacon's history.

Chapter 4
Beacon Crib Blankets

Millions of crib blankets were manufactured, but few remain today. Early crib blankets were available in all cotton, cotton-wool blends and all wool. Indian designs appeared in early catalogs, and the *Ombre* process was part of the baby blanket design scheme from the late 1920s, following much the same course as the regular blankets. In the early 1950s, both 100% cotton and blends composed of rayon and nylon were offered, and in 1955 100% Orlon was introduced.

Designer William H. Berner used playful characters for crib blanket designs that he adapted from children's coloring books.

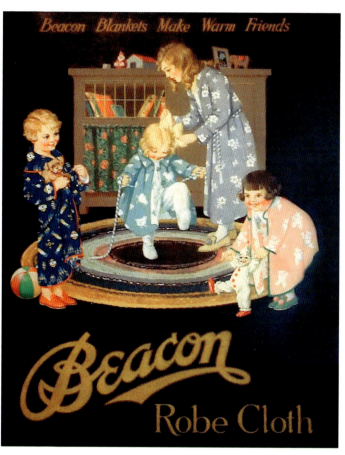

Beacon robe cloth, advertising painting. Courtesy of Beacon Manufacturing.

Crib blankets box cover, "c. 1920s." Courtesy of Dan Owenby.

"The Beacon Baby" adorned blanket packages and catalogs. Courtesy of Beacon Manufacturing Co.

A Beacon crib blanket and box in original condition. Courtesy of Beacon Manufacturing Co.

BEACON BLANKETS

Nursery designs and suitable plain and figured patterns in delicate pink and blue. These Crib Blankets may be carried profitably in both the Bedding and Infants' Departments

Advertising from 1927-28.

Beacon Ombre butterflies crib blanket, "c. 1930s." Courtesy of Sam and Denise Kennedy of Cisco's, Coeur d' Alene, Idaho.

Beacon Crib Blankets for 1930

G·BABY BLANKETS

STYLE G. Size 36x50. Bound around with six-inch Acetate Satin.
Boxed individually.

COLORS:

Blue   Pink

For plain colors in the above, see pages 19 and 20.

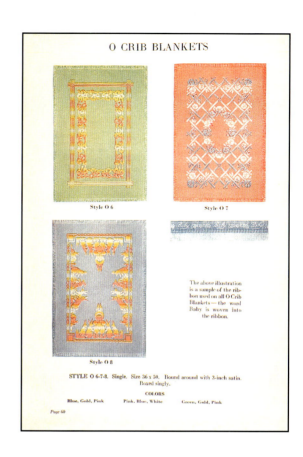

O CRIB BLANKETS

Style O 6   Style O 7

Style O 8

The above illustration
is a sample of the rib-
bon used on all O Crib
Blankets — the word
Baby is woven into
the ribbon.

STYLE O 6-7-8. Single. Size 36 x 50. Bound around with 3-inch satin.
Boxed singly.

COLORS

Blue, Gold, Pink   Pink, Blue, White   Green, Gold, Pink

STYLE C-256 — SINGLE — SIZE 36 x 50 inches
Bound around with 3-inch satin. Individually boxed.

COLORS:

Blue and White   Pink and White

J. C. PENNEY COMPANY, INC.

IX CRIB BLANKETS

Style IX   Style IX

Style IX   Style IX

STYLE IX. Single. Size 36 x 50. Stitched or bound all around with
2-inch sateen. Packed one dozen in a package only. Assorted designs;
6 of each color.

COLORS

Blue and White   Pink and White

FRINGED CARRIAGE ROBES

STYLE CR 4

STYLE CR 4. Single. Size 30 x 40. Fringed around. Boxed.

COLORS

Brown and Taupe 1   Blue and Taupe 3
Red and Taupe 2   Green and Taupe 4

STYLE CR 2. Single. Size 30 x 40. Fringed around. Boxed.

COLORS

Blue   Pink   White   Tan

CRIB BLANKET CATALOG 1932

● The 1932 catalog of Beacon Crib Blankets contains our full line of crib blankets, including new patterns and styles never before shown.

● The 1932 line has been carefully planned to meet 1932 consumer requirements. The price range is from the inexpensive Style S line to the more expensive Style F and LG blankets and the fringed carriage robes.

● Beacon Crib Blankets and juvenile robe cloth shown in this catalog are made in our mill at New Bedford, Massachusetts.

● Please note that this catalog covers only our Crib Blanket line and does not include our full line of Plaid, Reversible, solid-color, and Indian design blankets for general home use made at New Bedford or in our mill at Swannanoa, N. C. For our full line catalog, write to our nearest sales office (New York or Chicago) or the New Bedford mill.

● When ordering blankets remember that all blankets are shipped F. O. B. mill in which they are made.

● Visit our New York or Chicago display salesrooms on your next buying trip. If you are not on our regular mailing list, send your name and address to the Sales Promotion Manager, Beacon Manufacturing Company, New Bedford, Mass.

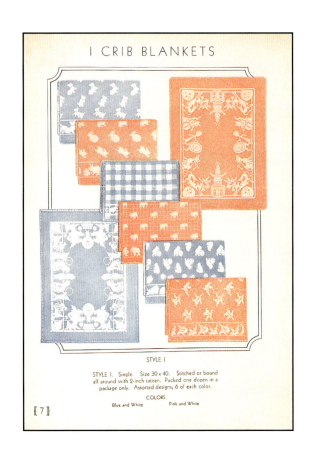

I CRIB BLANKETS

STYLE I

STYLE I. Single. Size 30 x 40. Stitched or bound all around with 2-inch sateen. Packed one dozen in a package only. Assorted designs; 6 of each color.

COLORS

Blue and White   Pink and White

CHILD'S BED BLANKETS

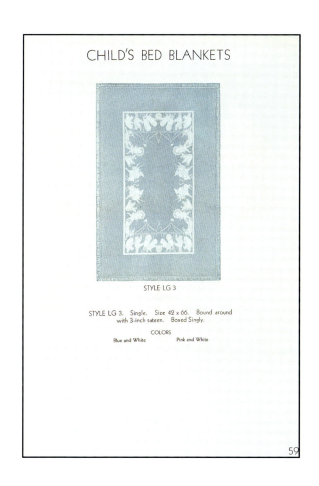

STYLE LG 3

STYLE LG 3. Single. Size 42 x 66. Bound around
with 3-inch sateen. Boxed Singly.

COLORS

Blue and White   Pink and White

O CRIB BLANKETS

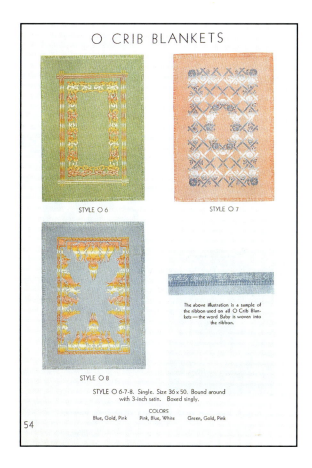

STYLE O 6   STYLE O 7

STYLE O 8

The above illustration is a sample of
the ribbon used on all O Crib Blankets — the word Baby is woven into
the ribbon.

STYLE O 6-7-8. Single. Size 36 x 50. Bound around
with 3-inch satin. Boxed singly.

COLORS

Blue, Gold, Pink   Pink, Blue, White   Green, Gold, Pink

54

F CRIB BLANKETS

STYLE F76   STYLE F81

STYLE F82   STYLE F83

STYLE F. Single. Size 36 x 50. Bound around with
3-inch satin. Boxed singly.

COLORS

Blue and White   Pink and White

(Page 15)

59

145

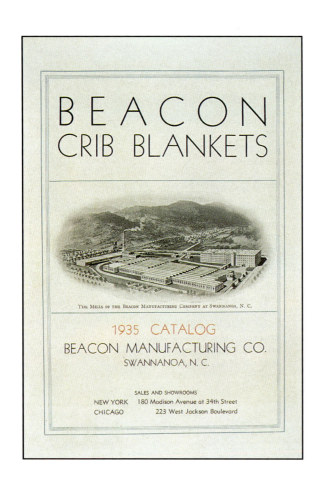

BEACON
CRIB BLANKETS

The Mills of the Beacon Manufacturing Company at Swannanoa, N. C.

1935 CATALOG
BEACON MANUFACTURING CO.
SWANNANOA, N. C.

SALES AND SHOWROOMS
NEW YORK   180 Madison Avenue at 34th Street
CHICAGO   223 West Jackson Boulevard

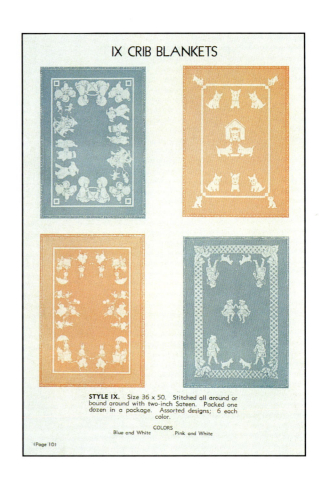

IX CRIB BLANKETS

STYLE IX.   Size 36 x 50.   Stitched all around or bound around with two-inch Sateen.   Packed one dozen in a package.   Assorted designs; 6 each color.

COLORS
Blue and White   Pink and White

(Page 10)

G CRIB BLANKETS

G 179   G 185

G 184   G 186

STYLE G.   Size 36 x 50.   Bound around with three-inch Rayon Binding.   Boxed Individually.

COLORS
Pink and White   Blue and White

(Page 12)

Beacon
BLANKETS
make warm
friends

BEACON
CRIB
BLANKETS
1938

BEACON MANUFACTURING COMPANY

G CRIB BLANKETS

STYLE G. Size 36 x 50. Bound around with four-inch
Rayolite. Boxed Individually.

COLORS:

Blue   Pink

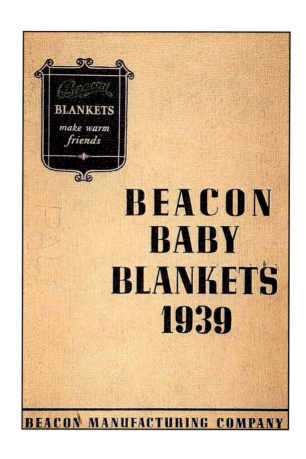

BLANKETS
*make warm
friends*

BEACON BABY BLANKETS 1939

BEACON MANUFACTURING COMPANY

A CRIB BLANKETS

STYLE A. Size 30 x 36. Stitched around.
One dozen in a package. Assorted designs; 6 each color.

COLORS:

Blue   Pink

F BABY BLANKETS

STYLE F. Size 36 x 50. Bound around with five-inch
Acetate Satin. Boxed Individually.

COLORS:

Blue   Pink

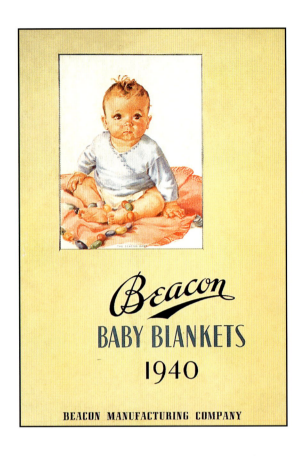

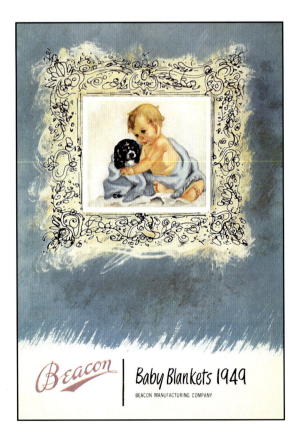

Seven examples from the 1953 Baby Catalog, all displaying William H. Berner's art. Courtesy of Owen Manufacturing Co.

F-216   F-217

F-224   F-225

STYLE F JACQUARD   36" x 50"   Bound around 7" Acetate Satin, Boxed
Packed 24 to carton
PATTERN 216 — Blue, White, Canary combination   Bound Blue
Pink, White, Nile combination   Bound Pink
Nile, White, Pink combination   Bound Nile
Canary, Nile, White combination   Bound Canary
PATTERN 217 — Blue and White, Pink and White, Nile and White, Canary and Blue
PATTERN 224 — Blue and White, Pink and White, Nile and White, Canary and Blue
PATTERN 225 — Blue and White, Pink and White, Nile and White, Canary and Blue
Also made in Solid colors as STYLE F1 — Blue, Pink, Canary, Nile, White

(30)

SANDMAN 188   SANDMAN 205

SANDMAN 210   SANDMAN 207

STYLE SANDMAN   30" x 40"   Stitched around
Packed 48 to carton Assorted Patterns and Colors
PATTERN 188 — Blue and White, Pink and White, Nile and White, Canary and Blue
PATTERN 205 — Blue and Pink, Pink and Blue, Nile and White, Canary and Nile
PATTERN 207 — Blue and White, Pink and White, Nile and White, Canary and Blue
PATTERN 210 — White, Blue, Pink, Canary combination
White, Blue, Pink, Nile combination

(23)

CAMEO 14   CAMEO 15

CAMEO 16

STYLE CAMEO   36" x 50"   Bound around 5" Acetate Satin, Boxed
Packed 24 to carton Assorted Patterns and Colors
PATTERN 14 — White, Blue, Pink, Nile combination   Bound Nile
White, Blue, Pink, Canary combination   Bound Canary
PATTERN 15 — Blue and White, Pink and White, Nile and White, Canary and Blue
PATTERN 16 — Blue and White, Pink and White, Nile and White, Canary and Blue

(27)

STARLING 308   STARLING 312

STARLING 323   STARLING 324

STYLE STARLING   36" x 50"   Bound around 3" Acetate Satin, Boxed
Packed 24 to carton Assorted Patterns and Colors
PATTERN 308 — Blue and White, Pink and White, Nile and White, Canary and Blue
PATTERN 312 — Blue, White, Pink, Canary combination   Bound Blue
Pink, White, Blue, Canary combination   Bound Pink
PATTERN 323 — Blue and White, Pink and White, Nile and White, Canary and Blue
PATTERN 324 — White, Blue, Pink, Canary combination   Bound Canary
White, Blue, Pink, Nile combination   Bound Nile
Also made in Solid Colors as STYLE STARLING No. 1 — Blue, Pink, Canary, Nile, White

(25)

IXA 324   IXA 325

STYLE 1 X A   36" x 50"   Bound around 3" Acetate Satin, Boxed
Packed 24 to carton Assorted Patterns and Colors
PATTERN 324 — Blue and White, Pink and White, Nile and White, Canary and Blue
PATTERN 325 — White, Blue, Pink, Canary combination   Bound Canary
White, Blue, Pink, Nile combination   Bound Nile

(26)

Chapter 5
Beacon Robe Fabrics

Beacon produced the bathrobe fabric, not the actual bathrobe. Fabric was manufactured in 27" and 36" width cloth to accommodate both children and adult sizes. The fabric was cut and sewn into bathrobes in New York by several different contractors.

Beacon "robing", as the fabric was known, was developed by Charles Owen Dexter. It had a tight short nap and was basically a double cloth composed of one set of warps and two sets of filling. Beacon made boxed robe kits for the home sewing consumer complete with instructions, neck cord and girdle set (waist cord). Girdle sets were also sold separately.

Robing fabric was an extremely successful product for Beacon. In 1923 there were 86 patterns available in plaids, geometric designs, floral and Indian designs. By 1929 the *Ombre* process (effect) was added and the number of patterns was consolidated to eighteen. Cotton Beacon robing was changed to a rayon blended fabric in the early 1950's and within ten years robing fabric was dropped from the line. Today Beacon vintage cotton robes are collectible.

Over the years, Beacon robes and blankets have appeared in numerous movies, and on a variety of television shows. Have fun looking for them.

Bathrobes box cover, "c. 1920s." Courtesy of Dan Owenby.

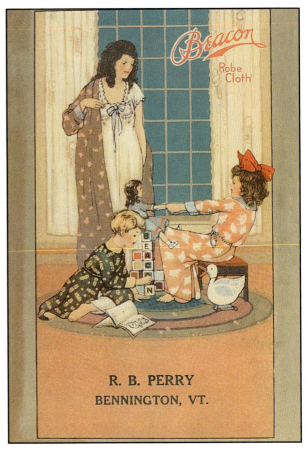

Robe cloth box cover, "c. 1920s." Courtesy of Dan Owenby.

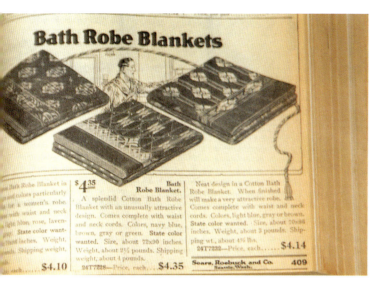

Bath Robe Blankets

1919 Bath Robe Blankets: Center item is described as a "Splendid Cotton Bath Robe blanket." Comes complete with neck and waist cords. $4.35 each. Courtesy of Sears, Roebuck & Co. archives, reprinted by special arrangement and protected by copyright. No duplication is permitted.

Wardrobe closets with beautiful Beacon blankets and a robe. Courtesy of Sam and Denise Kennedy of Cisco's, Coeur d'Alene, Idaho.

Beacon Ombre plaid robe, originally buttoned and tied. Courtesy of Charles D. Owen III.

Left: Page from Spring and Summer 1948 Montgomery Ward. Courtesy of Montgomery Ward Collection, American Heritage Center, University of Wyoming.

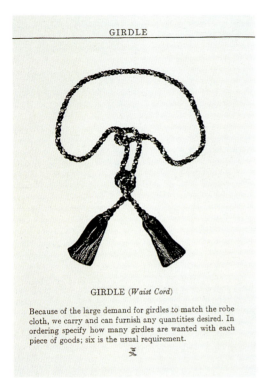

Girdle or waist cords were available from Beacon.

Trio of robes from the 1920s, in excellent condition. All three button and tie with original girdles. Courtesy of Sam and Denise Kennedy of Cisco's, Coeur d'Alene, Idaho.

1920 Ladies Robes. Top center, is a basic brown plaid, below is a blue Indian design. Between $4.00 and $7.00 each. Courtesy of Sears, Roebuck & Co. archives, reprinted by special arrangement and protected by copyright. No duplication is permitted.

Very early Beacon bathrobe worn by Robert Barone, photographed at the Laura Fisher/Antique Quilts and Americana store in New York.

Early Art Deco design robe, excellent condition with its original girdle and buttons, "c. 1920." Author's collection. $50-125.

1926 Ladies Robes. Lower left is a Beacon blanket cloth trimmed in quilted satin, $7.79 each. Top right, (in the fold), economy model, $4.39 each. Courtesy of Sears, Roebuck & Co. archives, reprinted by special arrangement and protected by copyright. No duplication is permitted.

Ombre robe, home assembled, featuring cuffs, matching fabric belt and button front, "c. 1930." Author's collection.

Ladies robe floral design, "c. 1930s." Courtesy of Beacon Manufacturing Co.

1935/1936 Men's Robes. Left, robe A, "Whittenton", Ombre design, $2.59 each. Right, robe C, Heavyweight Beacon blanket cotton cloth, $3.79 each. Courtesy of Sears, Roebuck & Co. archives, reprinted by special arrangement and protected by copyright. No duplication is permitted.

1936/1937 Ladies Robe, "Jubilee Special", Genuine Beacon Ombre, $2.79 each. Courtesy of Sears, Roebuck & Co. archives, reprinted by special arrangement and protected by copyright. No duplication is permitted.

1939/1940 Men's Robes. Left to right, "Big burly and warm, Heavyweight", $3.79; our best Beacon cloth, Ombre, $4.98 each. "Whittenton" heavy cotton (a house brand made by Beacon) $2.59 each. Courtesy of Sears, Roebuck & Co. archives, reprinted by special arrangement and protected by copyright. No duplication is permitted.

Excellent example of a Beacon robe from the late-1940s, original mint condition. Courtesy of Beacon Manufacturing Co.

Glen Plaid Beacon Robe, $7.90. Courtesy of J.C. Penny Company, Inc., Plano, Texas.

Maroon classic bathrobe, late-1940s. Courtesy of Beacon Manufacturing Co.

Glen Plaid Robe sold by J.C. Penny's in 1952. Courtesy of Beacon Manufacturing Co.

Chapter 6
Beacon Labels

Most Beacon blankets came with a removable paper label. Blankets bound on all four sides, blankets containing wool, and shawls were more likely to be labeled with a sewn-on fabric label. Bathrobes were generally labeled "Genuine Beacon Fabric" or "Made of Beacon Blanket" by the contractors who made them.

Early Beacon blankets sold either through Sears, Roebuck & Co or Montgomery Ward may have a label that reads "Indian Blanket Corporation, Chicago, Illinois".

The Esmond brand was acquired by Beacon in the late 1940s. Esmond had operated two plants, one in Rhode Island which closed down and another in Quebec that Beacon took over and operated. Esmond had produced cotton blankets from the 1920s to the 1940s. The Esmond sewn-on rabbit label symbolized "soft and warm as rabbit fur" and they also used a paper label.

There were several companies that produced cotton blankets and labeled them with their own brand names using paper or fabric labels. These included Chatham; Jack Frost; NEPAH' WIN (Spirit of Sleep) by Canadian Cotton Ltd; and Montauk, by Fair Trading Co. Inc.

"New, old stock" labeled blanket; Swannanoa plant appears in the background, "c. 1924." Courtesy of Laura Fisher/Antique Quilts and Americana.

Traveling rug label for a heavy cotton blanket, usually in plaid, for auto or picnic use, "c. 1924." Courtesy of Laura Fisher/Antique Quilts and Americana.

Wigwam paper label with New Bedford mill in the background, "c. 1910/1911." Courtesy of Charles D. Owen, Jr. IV.

Standard paper label used on the majority of Beacon products. Courtesy of Laura Fisher/Antique Quilts and Americana.

The "Esmond Warm Spun Blanket" label. A bunny appears on the reverse side. Courtesy of Charles D. Owen III.

Indian Blanket Corporation label appears on an Ombre shawl label by Beacon for either Sears, Roebuck & Co., or Montgomery Ward, both based in Chicago. Courtesy of Bearwallow Mountain Traders, Judy Judson & Norwood Barnes.

The Esmond bunny symbolizing, "soft and warm as rabbit fur". Beacon acquired Esmond in the late-1940s. Courtesy of Charles D. Owen III.

Blankets bound on all four sides were generally labeled. Courtesy of Sam and Denise Kennedy of Cisco's, Coeur d' Alene, Idaho.

NEPAH'WIN (Spirit of Sleep) label by Canadian Cotton Ltd. Author's collection.

Labels had just the one word on them, "Beacon" usually in gold on a black background. Courtesy of Gary Holt/Steve Christianson, Gary Holt Collection.

Label placement was in the lower right hand corner on the patternization side of a Beacon blanket. Courtesy of Charles D. Owen III.

"Genuine Beacon Fabric" robe label used by one of the New York contractors that manufactured Beacon bathrobes from Beacon robe cloth.

Fringed blankets or shawls were labeled in the lower right hand corner. Courtesy of Sam and Denise Kennedy of Cisco's, Coeur d'Alene, Idaho.

"Made of Genuine Beacon Blanket" An early example of a robe label. Author's collection.

The background colors of the label and the script have been changed to match the shawl. Courtesy of Sam and Denise Kennedy of Cisco's, Coeur d'Alene, Idaho.

The Casco (design) blanket label "c. 1990." Courtesy of David H. Schutzler.

Chapter 7
Other Related Blankets

Novel Blankets

Indian Chief design blanket, patternization side, perhaps a Beacon, "c. 1910." Author's collection. $175-300.

Colorization side of the same blanket. Author's collection.

Gray background Indian Chief design. Courtesy of Sam and Denise Kennedy of Cisco's, Coeur d'Alene, Idaho.

A blazing sun above teepees, outdoor log fires and Indian braves on horseback, probably a Beacon blanket, patternization side, early-teens, Author's collection. $300-500.

Indian warriors on horseback, possibly a Beacon blanket, "c. 1912." Courtesy of Sam and Denise Kennedy of Cisco's, Coeur d'Alene, Idaho.

Colorization side of blazing suns.

Teepees and Indian motifs combine on this early example of a manufactured Indian style blanket, early-teens, Author's collection. $175-300.

Left and below: The western design on this blanket runs from top to bottom with ropes and brands at each end. This is actually a mural on a blanket, manufacturer unknown, "c. 1940." Author's collection. $100-175.

The western design blanket in red and blue colorations, "c. 1940s." Courtesy of Sam and Denise Kennedy of Cisco's, Coeur d'Alene, Idaho.

Shrine for Democracy blanket, manufacturer unknown. Mount Rushmore opened October 31, 1941. Courtesy of Bearwallow Mountain Traders, Judy Hudson & Norwood Barnes.

Nautical blankets were popular for boys' rooms. This whaling design is unusually graphic, "c. 1940s." Author's collection. $40-75.

"God Bless America," and eagles are located at each end of this patriotic blanket with red, dazzling stars in the blue center field, "c. 1940s." Courtesy of Laura Fisher/Antique quilts and Americana.

Out of this world! Spacemen and flying saucers, great dream material on a child's blanket, "c. 1950s." Courtesy of Sam and Denise Kennedy of Cisco's, Coeur d'Alene, Idaho.

Hunting scene blanket, manufacturer unknown, "c. 1950s." Courtesy of Sam and Denise Kennedy of Cisco's, Coeur d'Alene, Idaho.

A Beacon design blanket or a blanket with a Beacon design? J.C. Penny originally sold similar type blankets for $3.69 in 1948. Author's collection. $60-95.

Institutional Blankets

Great Northern Railway, 100% wool blanket in a small brown and white checkered pattern, "c. 1920s." Author's collection. $75-275.

Close-up of the great Northern Railway logo and blanket pattern, "c. 1920s." Author's collection.

Great Northern Railway train car and blanket at the Issaquah, Washington train station. Author's collection.

1928 catalog page shows samples of hotel and organization wool blankets.

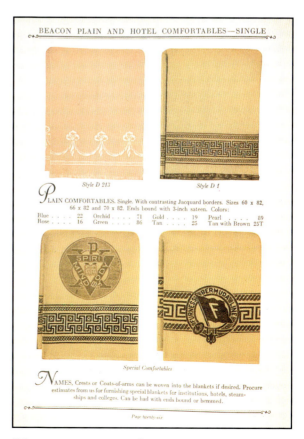

"Names, crests or coats of arms can be woven into the blankets if desired", states the caption from the 1928 Beacon catalog.

The Car and The Blanket

A splendid 1937 Chevrolet Master, owned by Dave and Connie Filkins of Puyallup, Washington, is complete with a Beacon blanket of the same vintage. The blanket, a version of a Signet or Midway design, has been washed countless times since it was purchased by Connie's mother around 1938 at their local JC Penny's store. It hangs behind the front seat on the grab bar ready for picnics, beach outings, or for use as a car robe.

The Rustic Lodge Look

In the beautiful, rolling, green hills near the Beacon plant in Swannanoa, a lodge remains with rustic cabins from a bygone era. Built in 1926, this wonderful structure is an antique treasure, still welcoming guests.   Vintage Beacon blankets were used here, purchased at the Beacon Factory Outlet store a few miles away. What a delightful find for the unsuspecting traveler.

A vintage Beacon blanket displayed on an original railing made from rhododendron branches, "c. 1950s." Courtesy of Carolyn A. Bartlett, Pine Lodge Rustic Cabins, Black Mountain, North Carolina.

This rustic cabin comes complete with vintage Beacon blankets on the bed, "c. 1950s." Courtesy of Carolyn A. Bartlett, Pine Lodge Rustic Cabins, Black Mountain, North Carolina.

Beacon blankets and shawls make a stunning impression draped over rustic chairs, on a lawn swing, or when used as wall hangings. Blankets "c. 1910-1930." Courtesy of Brenda Cain, Santa Monica, California, brendacain.com

Bully Good (Skookum)
Native American Dolls

Skookum means "bully good" in Siwash. Starting in 1913 in Missoula, Montana, Mary Dwyer McAboy used dried apples for the heads of her dolls. She pinched out the nose, lips and eye sockets and used bead-headed pins for the eyes, and best of all, wrapped her dolls in pieces of Beacon blankets.

Her early Indian dolls were fashioned after Salish and Kootenai people living on the Flathead Indian Reservation near her home. Later she developed her creations to reflect a variety of different Native Americans.

McAboy traveled extensively about the Indian reservations, attending Pow Wows and modeling clay heads of handsome and important chiefs. She saw Native Americans wearing Beacon blankets every day and adopted these blankets for her dolls because, as McAboy stated, "they were easily available, inexpensive and symbolized true Native American design."

Karen B. Kurtz, author of an article in *Doll World Magazine*, April, 1994, explains, "All dolls were wrapped in scraps of cotton Beacon blankets which are tightly folded over their bodies to simulate arms, as arms and hands are nonexistent." Commercial manufacturing of Skookum dolls was begun in 1917 by George Borgfeldt and the dolls were distributed by Arrow Novelty Company in New York. Later H. H. Tammen Company of Denver, Colorado, also produced and distributed Skookums.

An outstanding collection of Skookum dolls all in various Beacon style blankets. Courtesy of Sam and Denise Kennedy of Cisco's, Coeur d'Alene, Idaho.

Skookum Indian family with a baby peeking over its mother's shoulder. The dolls are wearing Beacon style blankets. Courtesy of Marvin Leib. $65-285.

169

Two baby Skookum dolls wrapped in Beacon style blankets and with their mailing labels attached. They could be mailed as souvenirs from the Southwest. Courtesy of Marvin Leib. $25-50.

Skookum Indian chief and his bride wearing Beacon style blankets on display among Indian-motif china. Courtesy of Sierra Hills Antiques, Grass Valley, California.

Chapter 8

Prices: Then and Now

The Beacon 1928 Price List

The reference to "Camp Blankets" (instead of camping blankets) may be the origin of the present-day term.

Beacon Wholesale Price List, May 1st, 1928.

NOTE CHANGE IN NEW YORK OFFICE ADDRESS

Price List, May 1st, 1928

PRICE PROTECTION GUARANTEED—If Prices are reduced, all goods unshipped will be billed at the new prices.

Part Wool Plain Blankets (Pairs)

Style XN	60 x 80, bound with 3-inch Sateen ....per pair	$2.60
" X	66 x 80, " " " " ...." "	2.80
" XW	72 x 84, " " " " ...." "	3.35

In white, gray and tan

Part Wool Blankets (Pairs)

Style B 70 x 80, bound with 4-inch sateen..............per pair $3.15
Colors: Blue, rose, orchid, green, and gold with white borders, also tan with brown border (2518).

Part Wool Plaid Blankets (Pairs)

In a large variety of designs and colorings.
Style H 70 x 80, bound with 4-inch sateen..............per pair $3.15
" P 70 x 80, " " " " ...." " 4.00

Part Wool Blankets (Single)

Priscilla 66 x 80, bound with 3-inch sateen..............each. $2.15
Colors: Blue, rose, orchid, green, and gold with white borders, also tan with brown border (2518).

Part Wool Blankets, Jacquard Borders (Single)

Diana 66 x 80, bound with 3-inch sateen..............each $2.35

Part Wool Plaid Blankets (Single)

In a large variety of designs and colorings.
Signet 66 x 80, bound with 3-inch sateen..............each $2.15
Topaz 70 x 80, " " " " ...." 2.75
Yukon 66 x 84, " " " " ...." 3.35
" 72 x 84, " " " " ...." 3.65

Part Wool Camp Blankets (Single)

Camping Blanket, Style CB, 60 x 82..............each $2.35
Above made in
Khaki mixture with dark khaki border.
Dark blue mixture with black border.

Indian Blankets

Agawam 60 x 80, bound all around ....each, boxed $2.35
Wigwam 66 x 80, " " " Part Wool.............. 3.00

Part Wool Ombre Indian Blankets

Huron 66 x 80, bound all around..............each, boxed $3.75

Part Wool Ombre Blankets

A heavy top blanket with silky rainbow effect obtained by the shading of rich colors.
Style CO 66 x 80, bound with 3-inch sateen..............each, boxed $3.75

NOTE CHANGE IN NEW YORK OFFICE ADDRESS

Price List, May 1st, 1928

PRICE PROTECTION GUARANTEED—If Prices are reduced, all goods unshipped will be billed at the new prices.

Part Wool Plain Blankets (Pairs)

			See Catalog Page
Style XN	60 x 80, bound with 3 inch Sateen ....per pair	$2.60	6
" X	66 x 80, " " " " ...." "	2.80	6
" XW	72 x 84, " " " " ...." "	3.35	6

In white, gray and tan

Part Wool Blankets (Pairs)

Page
Style B 70 x 80, bound with 4-inch sateen..............per pair $3.15 7
Colors: Blue, rose, orchid, green, and gold with white borders, also tan with brown border (2518).

Part Wool Plaid Blankets (Pairs)

Pages
In a large variety of designs and colorings.
Style H 70 x 80, bound with 4-inch sateen..............per pair $3.15 8–9
" P 70 x 80, " " " " ...." " 4.00 10–11

Part Wool Blankets (Single)

Page
Priscilla 66 x 80, bound with 3-inch sateen..............each. $2.15 12
Colors: Blue, rose, orchid, green, and gold with white borders, also tan with brown border (2518).

Part Wool Blankets, Jacquard Borders (Single)

Page
Diana 66 x 80, bound with 3-inch sateen..............each $2.35 14

Part Wool Plaid Blankets (Single)

Pages
In a large variety of designs and colorings.
Signet 66 x 80, bound with 3-inch sateen..............each $2.15 13
Topaz 70 x 80, " " " " ...." 2.75 16–17
Yukon 66 x 84, " " " " ...." 3.35 18–20
" 72 x 84, " " " " ...." 3.65 18–20

Part Wool Camp Blankets (Single)

Page
Camping Blanket, Style CB, 60 x 82..............each $2.35 6
Above made in
Khaki mixture with dark khaki border.
Dark blue mixture with black border.

Indian Blankets

Pages
Agawam 60 x 80, bound all around ....each, boxed $2.35 28–29
Wigwam 66 x 80, " " " Part Wool.............. 3.00 30–31

Part Wool Ombre Indian Blankets

Page
Huron 66 x 80, bound all around..............each, boxed $3.75 32

Part Wool Ombre Blankets

Pages
A heavy top blanket with silky rainbow effect obtained by the shading of rich colors.
Style CO 66 x 80, bound with 3-inch sateen..............each, boxed $3.75 20–21

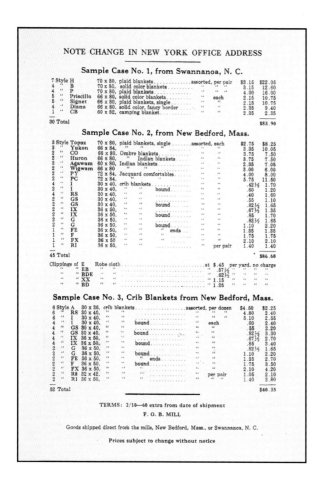

Evaluating a Beacon Blanket Today

Key factors in making an evaluation of a Beacon blanket today are condition, age, design, rarity and the blanket's or robe's fundamental appeal.

The *Ombre* blanket and *Ombre* robe cloth made their debuts in the mid-1920s and added greatly to the attractiveness of Beacon's products. The *Ombre* effect has also added greatly to today's collectibility and price.

Beacon Indian design blankets currently sell from approximately $60 to $750 or more, with a mint condition blanket obviously bringing the higher price. If a blanket has a yarn fringe, it is considered a shawl which can add or subtract from the price depending on its desirability, condition, color, and design.

Wool and cotton blend Indian design blankets in good condition fall into this same price range.

Plaid blankets are currently valued between $25 and $100, depending on condition, and an *Ombre* plaid can bring from $50 to $200.

If a blanket or shawl is labeled its value increases by an additional $25 to $75.

"Blanket pairs", the name given to blankets that are two blanket "lengths", can be somewhat confusing. The *Sears 1916 Fall Catalog* described them this way: "What a Blanket Pair Means: The word pair means two blankets but they do not come separate. They are woven in one continuous length. The size given is for one blanket and the pair of continuous length you receive will be twice the size given." These blanket pairs were available in cotton or wool blend, usually in two color plaid design, and bring $50 to $150 today.

Wool institutional blankets are valued at $75 to $275.

Bathrobes, adult sizes, sell for $35 to $125, with *Ombre* patterns being more valuable.

In 1990 Beacon produced a retro blanket line for L.L. Bean Company using the vintage Casco blanket design and an eighty-five percent wool, fifteen percent cotton blend. These are labeled, "Beacon Blankets Make Warm Friends. Dry Clean Only". Their packaging was labeled either Beacon Heritage or Beacon Trading Post. These blankets are currently valued from $75 to $175.

Beacon At The Century Mark

The Beacon Manufacturing Co., will turn one hundred in the year 2004. Among its customers have been Ralph Lauren, K-Mart Corporation, Sears, Roebuck Co., J.C. Penny, Montgomery Ward, Wal-Mart and the health care and hospitality industries. It has been estimated that Beacon has manufactured over a billion blankets since its founding in 1904.

Beacon produced the Ralph Lauren™ blanket line until recently. Designs for the line were taken from vintage Beacon blankets and although it was impossible to recreate the Ombre look, it was simulated.

Ralph Lauren™ Indian design blankets after cleaning, waiting for further processing.

Ralph Lauren™ throws being woven at Beacon, Swannanoa, 1996.

Walt Disney™ throws being woven at Beacon, Swannanoa, 1996.

U-Haul™ non-woven moving pads coming down the production line at Beacon, 1996.

Resources

Following is a partial list of dealers in vintage Beacon blankets that are known to the authors at the time of writing. No preference is implied and there are many other dealers handling period blankets.

Judy Hudson and Norwood Barns
Bearwallow Mountain Traders
919 Main Street Highway 64/74
P.O. Box 218
Chimney Rock, NC 28720
828-625-0996 or 828-628-0905

Brenda Cain
brendacain.com

Sam & Denis Kennedy
Cisco's
212 North 4th Street
Coeur d'Alene, ID 83814
208-769-7575

Jackie Ericson
Cellar 105 Antiques
105 West Main Street
Bozeman, MT 59715
406-587-3013

Laura Fisher/Antique Quilts and Americana
1050 2nd Ave.
Gallery 84
New York, NY 10022
212-838-2596

Gary Holt/Steve Christianson
Gary Holt Collection
248A North Higgins #422
Missoula, MT 59802
206-619-4813

David W. Schutzler
P.O. Box 1064
Enumclaw, WA 98022
360-825-0859

Jane Ross
Sierra Hills Antiques
P.O. Box 639
Cedar Ridge, CA 95924
530-273-9095

Paul and Stephanie Hauger
Yellowstone Vintage Clothing Co.
527 State Street
Santa Barbara, CA 93101
805-963-9609
and
712 North LA Brea Av.
Hollywood, CA 90038
323-931-6616

Bibliography

Obituaries

Dexter, Charles O. *Daily News Record*, New Bedford, Massachusetts, April 16, 1943.

Owen, Catherine. *Asheville Citizen Times*, Asheville, North Carolina, September 23, 2000.

Owen, Charles D. Sr. *Asheville Citizen Times*, Asheville, North Carolina, May 25, 1937.

Publications

Bell, Lisa. *A View From The Top*. Asheville: Montgomery Alabama Community Communications, Inc. 1996.

Carpo, Henry H. *The Story of Cotton and Its Manufacture Into Cloth,* paper read at the meeting of the Old Dartmouth Historical Society, New Bedford, Massachusetts, November, 1937.

Carter, Ted B. "A Family Success Story," *Black Mountain Newspaper*, March 29, 1977 and March 30, 1977.

"Cotton Manufactures of New Bedford," *Davison's Textile Blue Book*. United States and Canada. New York: Davison Publishing Co., 1914-1915.

Highsmith, William E. *The University of North Carolina at Asheville:The First Sixty Years*. Asheville, North Carolina: The University of North Carolina at Asheville, 1991.

"In the Matter of Beacon Manufacturing Company," *Federal Trade Commission Decisions*, June 28, 1932.

J.C. Penny Advertising: *Lebanon Express Newspaper* Lebanon, Oregon, August 11, 1946 and November 28, 1952.

Kurtz, Karen B. "Skookum's Bully Good Native American Dolls," *Doll World Magazine*, April 1990, pp. 5-6.

Montgomery Ward Catalog, Spring-Summer 1948, Montgomery Ward Collection, American Heritage Center, University of Wyoming, Laramie,WY.

Pease, Zephaniah W. *History of New Bedford, Massachusetts*. New York: The Lewis Historical Publishing Company, 1918.

Sears, Roebuck and Company Catalogs, 1919, 1920,1926,1935/1936, 1936/1937 and 1939/1940. Catalog pictures within this book are reprinted by arrangement with Sears, Roebuck and Company and are protected under copyright. No duplication is permitted.

Spanierman Gallery, LLC. New York City, biography of Norman Rockwell. AskART.com.

Stone, Orra L. *The History of Massachusetts Industries: Their Inception, Growth and Success*. vol. 1. Boston: S.J. Clark Publisher, 1930.

Sunday Standard Newspaper, New Bedford, MA: "*World Wide Respect For Beacon Blankets Won on Sheer Merit.*", p. 1, September 30, 1923.

"Textile," *Encyclopedia of Science and Technology*, Mc Graw Hill, 1992.

Van Vechten-Lindberry. *Biography of R(obert) Farrington Elwell, Artist Illustrator, Cowboy*. Taos, NM: Taos Art Museum. AskART.com

Wingate, Isabel B. *Textile Fabrics and Their Selection*. 5th ed. Englewood Cliffs, NJ, Prentice-Hall, Inc., 1964.

Woloshuk, Nicholas. *E. Irving Couse 1866-1936*, Santa Fe, NM: Santa Fe Village Art Museum, 1976.

Young, Marjorie W., ed. *Textile Leaders of The South*, Anderson, SC: J.R. Young (1963)